Primarily Plants

Authors
Carol S. Gossett
Evalyn Hoover
Kay Kent
Sheryl Mercier
Myrna Mitchell
Dinah Brown

Illustrator
Dawn DonDiego

Editors
Betty Cordel
Michelle Pauls

Desktop Publisher
Tanya Adams

This book contains materials developed by the AIMS Education Foundation in cooperation with the Fresno Unified School District.

This book contains materials developed by the AIMS Education Foundation. **AIMS** (**A**ctivities **I**ntegrating **M**athematics and **S**cience) began in 1981 with a grant from the National Science Foundation. The non-profit AIMS Education Foundation publishes hands-on instructional materials (books and the quarterly magazine) that integrate curricular disciplines such as mathematics, science, language arts, and social studies. The Foundation sponsors a national program of professional development through which educators may gain both an understanding of the AIMS philosophy and expertise in teaching by integrated, hands-on methods.

Copyright © 1990, 2005 by the AIMS Education Foundation

All rights reserved. No part of this work may be reproduced or transmitted in any form or by any means—graphic, electronic, or mechanical, including photocopying, taping, or information storage/retrieval systems—without written permission of the publisher unless such copying is expressly permitted by federal copyright law. The following are exceptions to the foregoing statements:

- A person or school purchasing this AIMS publication is hereby granted permission to make up to 200 copies of any portion of it, provided these copies will be used for educational purposes and only at that school site.

- An educator presenting at a conference or providing a professional development workshop is granted permission to make one copy of any portion of this publication for each participant, provided the total does not exceed five activities per workshop or conference session.

Schools, school districts, and other non-profit educational agencies may purchase unlimited duplication rights for AIMS activities for use at one or more school sites. Visit the AIMS website, www.aimsedu.org, for more information or contact AIMS:

• P.O. Box 8120, Fresno, CA 93747-8120 • 888.733.2467 • 559.255.6396 (fax) • aimsed@aimsedu.org •

ISBN: 1-932093-17-6

Printed in the United States of America

Primarily Plants

Table of Contents

Why Study Plants? 1

Science Information: Plant Growth 2
    Inside a Seed 3
    A Seed Grows 10
    A Plant Begins 18
    It's in the Bag 24
    Little Brown Seeds 30
    Soak and Sprout 35

Science Information: Seeds, Spores, and More 42
    Seed Sort 43
    The Seed Within 50
    Seed Soakers 57
    Seeds Travel 66
    Observing Bulbs 73
    Plants from Cuttings 79
    Spores, A Special Seed 83

Science Information: Plant Needs 87
    Which Soil Works Best? 88
    Plants and Water 93
    Blue Ribbon Crops 97
    Plants and Sunlight 105
    Plants and Space 109
    Patchwork Planting 113
    What do Plants Need to Grow? 120
    What Temperature is Best? 126
    What do Plants Need? 130
    Reaching up Toward the Sun 139
    People Need Plants 149

Science Information: Plant Parts 153
    Observe a Leaf 154
    Leafy Facts 158
    Leaf Safari 160
    Stem Study 166
    Super Tuber 173
    Root Study 179
    This is My Flower 184

Glossary 191

Why Study Plants?

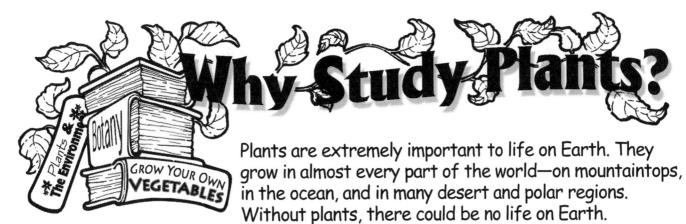

Plants are extremely important to life on Earth. They grow in almost every part of the world—on mountaintops, in the ocean, and in many desert and polar regions. Without plants, there could be no life on Earth. Plants produce the oxygen in the air, and the food we eat comes from plants or animals that eat plants.

Plants supply people with food, clothing, and shelter. Many of our most useful medicines are made from plants. Plants also provide us with beauty and pleasure.

Plants are probably most important to people as food. We eat seeds of plants, such as corn, rice, and wheat. When we eat carrots or beets, we are eating the roots of plants. We eat stems of asparagus and celery plants, the leaves of lettuce and cabbage plants, the flowers of broccoli and cauliflower plants, and the fruits of apple, banana, and orange trees.

Important raw materials come from plants. Trees give us lumber and other products. Cotton fiber is used for fabrics and clothing. Hemp is used to make rope.

Plants also provide an important source of fuel. People all over the world use wood to heat their homes or cook their food. Three sources of our fuel—coal, oil, and natural gas—all come from plants that lived long ago.

More than 60% of the drugs we have today come directly from or are derived from plants. Aspirin is a modification of salicylic acid that is found in the bark of willow trees. Penicillin comes from tiny plants called fungi. Yeast, a fungus, is used to produce alcohol and to make bread rise.

Plant Growth

Plants are organisms that grow and reproduce their own kind. They must have food, air, water, sunlight, and space in order to grow.

Green plants produce food and oxygen from water, carbon dioxide, and minerals through a process called photosynthesis. They take in carbon dioxide from the air, water and minerals from the soil, and energy from the sunlight. During photosynthesis, carbon dioxide and water are united in the presence of chlorophyll to form sugar and oxygen. The plant uses some of this food as it grows and produces leaves and fruit. The remaining food is converted to starch and stored in the plant.

Scientists have identified more than 350,000 kinds of plants. They fall into two basic categories—flowering and non-flowering plants. Those that produce flowers grow from seeds while non-flowering plants such as ferns, mosses, molds, and mildew grow from spores.

Flowering plants grow from seeds. A sprouting seed must absorb water before it will start to grow. It must also have soil firmly packed around it and warmth from the sun. Inside the seed is a tiny embryo surrounded by stored food. When the embryo starts growing, roots grow downward and a stem grows upward. Once the stem breaks through the surface of the soil into the sunlight, the first two true leaves form and the plant begins to make food. When plants have water, sunlight, and the proper minerals in the soil, they grow, manufacture food, and give off oxygen.

Non-flowering plants grow from spores. Like a seed, a spore develops into an embryo. Unlike a seed, the spore does not contain food for the embryo to grow. The plant that develops must get food elsewhere.

Inside a Seed

Topic
Seeds

Key Question
What does the inside of a seed look like?

Learning Goals
Students will:
- observe lima beans that have been soaked in water, and
- identify the major parts of the seed.

Guiding Documents
Project 2061 Benchmarks
- *Most things are made of parts.*
- *Describe and compare things in terms of number, shape, texture, size, weight, color, and motion.*

NRC Standard
- *Each plant or animal has different structures that serve different functions in growth, survival, and reproduction. For example, humans have distinct body structures for walking, holding, seeing, and talking.*

Math
Measurement
    length

Science
Life science
    botany
        seed structure

Integrated Processes
Observing
Comparing and contrasting
Recording data

Materials
Lima beans
Student pages

Background Information
Seeds consist of two major parts: the outside seed coat and the inside embryo and stored food. The seed coat protects the developing plant, the embryo. In a lima bean seed, there are two cotyledons. These store food for the embryo. They are the leaves that are attached to the embryo. When the seed begins to grow, one part of the embryo becomes the root and the rest becomes the upper stem and leaves.

Management
1. Large lima been seeds are easy to handle and the parts are easily distinguishable.
2. Soak the beans overnight so the seed coats are loose.

Procedure
1. Distribute one water-soaked seed and one dry seed to each student. Also distribute the student pages.
2. Have students observe and describe the dry seed. Encourage them to use terms of color, texture, firmness, etc.
3. Direct the students to measure the seed by tracing it on the ruler.
4. Invite them to compare (look at likenesses) and contrast (look at differences) the dry seed and the wet seed. List the similarities and differences on the board or overhead projector. Also have students record the differences on the first student page.
5. Have students look at the seed coat and find the spot where the seed was attached to the pod. Tell them that this small hole in the seed coat lets water into the seed.
6. Invite students to carefully remove the seed coat of the wet seed and place it to one side. Have them identify the seed coat on the second student page.
7. Have them split the wet seed into two parts. Ask them to share their observations. Have them identify the embryo (little plant) and record its location on the second student page.

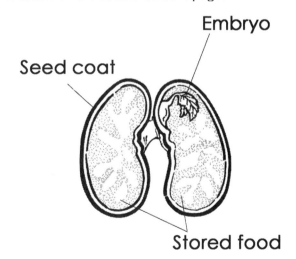

8. Point out that the remaining part of the seed is the food storage area of the plant. Tell them that the embryo gets its food from this area. Have them identify the food storage areas on their student page.

PRIMARILY PLANTS

Connecting Learning
1. What did the dry seed look like? Was the skin wrinkled? Could you find the spot on the outside of the seed? Why is that spot there?
2. How did the soaked seed differ from the dry seed?
3. What did the seed coat you removed look like?
4. Describe the insides of the seed.
5. Why is each part necessary?
6. What did the embryo look like? Can you see the shape of the future leaves?
7. How does the food storage area of your seed compare to the food storage area of another person's seed? Were they the same size? ...same shape? ...same color? ...same texture?
8. Do you think all seeds look alike when they are split open? Explain.
9. What are you wondering now?

Extensions
1. Use red beans to compare with the lima beans.
2. Try other large dicot seeds to see how they compare with the lima bean.
3. Open peanuts (roasted or plain). Observe the embryo inside the bean. Ask the students which parts of the peanut they are eating. [The embryo and the food storage. (Some peanuts still have the skins (the seed coat) on them.)]

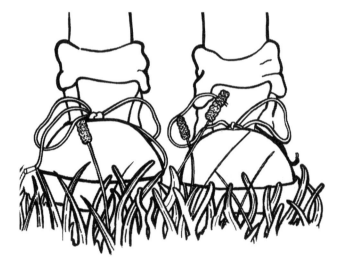

Inside a Seed

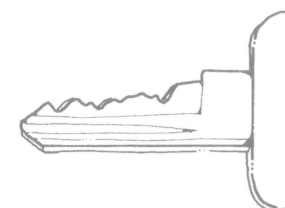

Key Question

What does the inside of a seed look like?

Learning Goals

Students will:

- observe lima beans that have been soaked in water, and
- identify the major parts of the seed.

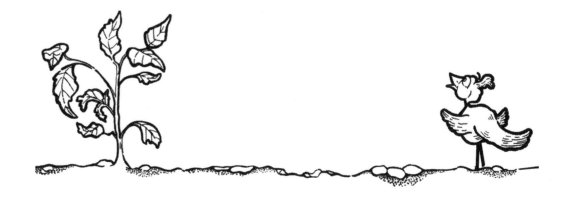

Inside a Seed

1. What does the dry seed look like?

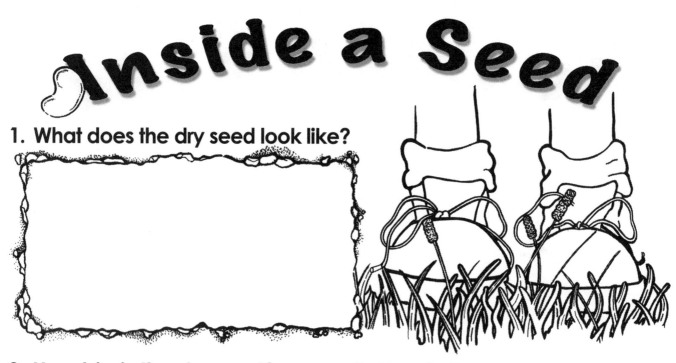

2. How big is the dry seed? Trace the seed on the ruler.

| 1 | 2 | 3 | 4 | 5 | 6 | 7 | 8 |

3. How is the wet seed different from a dry seed?

4. Split the seed in half. What does it look like on the inside? Look for the tiny plant called the embryo.

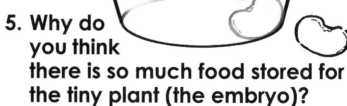

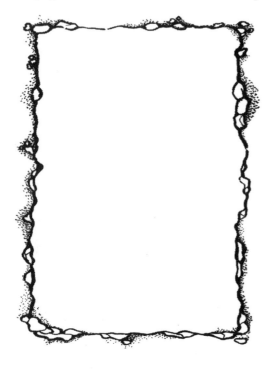

5. Why do you think there is so much food stored for the tiny plant (the embryo)?

PRIMARILY PLANTS © 2005 AIMS Education Foundation

Inside a Seed

Label the parts:
food storage, seed coat, little plant (embryo).

1. _____
2. _____
3. _____

PRIMARILY PLANTS

Inside a Seed

Connecting Learning

1. What did the dry seed look like? Was the skin wrinkled? Could you find the spot on the outside of the seed? Why is that spot there?

2. How did the soaked seed differ from the dry seed?

3. What did the seed coat you removed look like?

4. Describe the insides of the seed.

5. Why is each part necessary?

Inside a Seed

6. What did the embryo look like? Can you see the shape of the future leaves?

7. How does the food storage area of your seed compare to the food storage area of another person's seed? Were they the same size? ...same shape? ...same color? ...same texture?

8. Do you think all seeds look alike when they are split open? Explain.

9. What are you wondering now?

A Seed Grows

Topic
Plants

Key Question
How does a seed grow?

Learning Goal
Students will observe the growth of a bean seed.

Guiding Documents
Project 2061 Benchmarks
- Describing things as accurately as possible is important in science because it enables people to compare their observations with those of others.
- Plants and animals both need to take in water, and animals need to take in food. In addition, plants need light.
- Simple graphs can help to tell about observations.

NRC Standards
- Employ simple equipment and tools to gather data and extend the senses.
- Organisms have basic needs. For example, animals need air, water, and food; plants require air, water, nutrients, and light. Organisms can survive only in environments in which their needs can be met. The world has many different environments, and distinct environments support the life of different types of organisms.

NCTM Standards 2000*
- Recognize the attributes of length, volume, weight, area, and time
- Represent data using tables and graphs such as line plots, bar graphs, and line graphs
- Recognize and apply mathematics in contexts outside of mathematics

Math
Measurement
   linear
Graphing
   bar graph
Counting

Science
Life science
   botany
      plant growth

Integrated Processes
Observing
Comparing and contrasting
Collecting and recording data
Interpreting data
Generalizing

Materials
Lima beans
Zipper-type plastic bags, pint size
Paper towels
Transparent tape
Paper clips
Student pages

Background Information
Seeds start to grow when conditions are right to support the needs of growing plants. Water, air, and proper temperature are all necessary for seed growth. Water makes the seed swell and softens the seed coat. The embryo begins to grow. A warm temperature is also needed for a seed to germinate. Many seeds begin to grow in the springtime when the sun shines and the days begin to warm. After a seed starts to grow, the embryo grows into a young plant. Seedlings need warm temperatures, water, and food to keep growing

Management
1. Each student should make a plastic bag greenhouse.
2. To inhibit mold growth on the seeds, rinse seeds with a weak bleach solution (five milliliters of bleach in one liter of water). Have students wash their hands before and after planting seeds.
3. Copy the greenhouse onto green paper or let the children color it.
4. If hanging the greenhouses in a window, use painter's masking tape so there will not be any adhesive residue left on the windows.

Procedure
1. Ask the *Key Question* and state the *Learning Goal.*
2. Distribute a lima been seed, a paper towel, and a plastic bag to each student.
3. Have the students fold the paper towel into fourths so that it will fit into the plastic bag.
4. Give each student a an 8-cm (3-inch) length of tape with which to adhere the bean to the paper towel. Have them position the bean in the approximate center of the folded paper towel.
5. Tell them to dampen their paper towel and then slide it into the plastic bag. Have them seal their bags.

6. Distribute the copies of the greenhouse. Have students write their names on the greenhouse and cut out the middle section.
7. Staple the greenhouse picture onto the plastic bag so that the bean shows through the cutout.
8. Distribute one ruler and a paper clip to each student. Explain that the zero on the ruler is in the middle. Tell them that they will put the zero on the seed and use one set of numbers to measure the stem as it grows up; the other set of numbers will measure the root as it grows down. Have students attach the ruler to the side of the greenhouse using the paper clip.
9. Hang the greenhouses on a window or in an area where students can easily observe and measure the growth.
10. Have students write their predictions (guesses, at this point) on the student page. Each day, have students observe and record their measurements on their graphs.
11. Conclude by having students record their actual results, compare their actual results to their predictions, and discuss the information on their graphs.

Connecting Learning
1. Describe the growth of the roots and the stem.
2. How do the roots and stems compare in color and direction?
3. How long did it take for your seed to sprout? Did everyone's seeds sprout on the same day? Why do you think there were differences?
4. How did you provide for the needs of the seed for growth?
5. What are you wondering now?

Extensions
1. Compare other seeds to the lima bean seeds.
2. Construct a plant growth view box. Remove the top from a small milk carton Cut a window in the side of the carton. Cover the window with plastic wrap. Fill the carton with soil and plant the seed next to the window. Keep the flap closed and open only to view the seed growing.

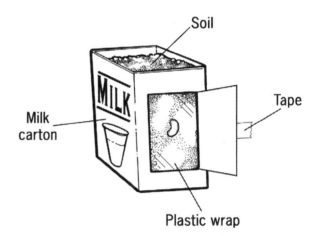

3. Use four beans. Position them so they face four different ways: up, down, left, and right. Tape them to a paper towel, dampen, and place in a plastic bag. Will the different orientation of the seeds affect the germination of the seeds?
4. Mix a cup of plaster of Paris with water. Pour into a Styrofoam cup. Place a dry lima bean in the middle of the mixture. When the plaster of Paris is dry, peel the cup from around it. Set it on a shelf and watch what happens. (The lima bean should swell from the moisture in the plaster of Paris and crack the plaster.)

* Reprinted with permission from *Principles and Standards for School Mathematics*, 2000 by the National Council of Teachers of Mathematics. All rights reserved.

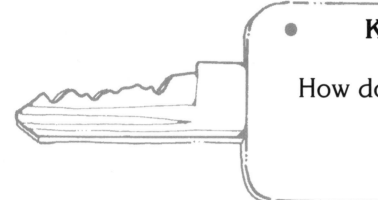

Key Question

How does a seed grow?

Learning Goal

Students will:

observe the growth of a bean seed.

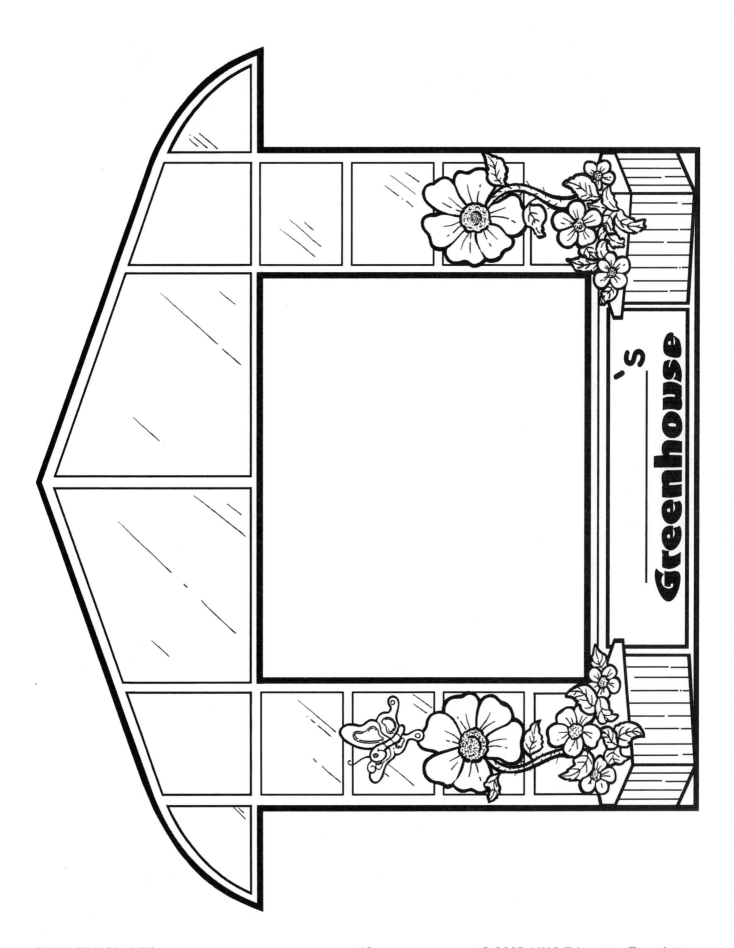

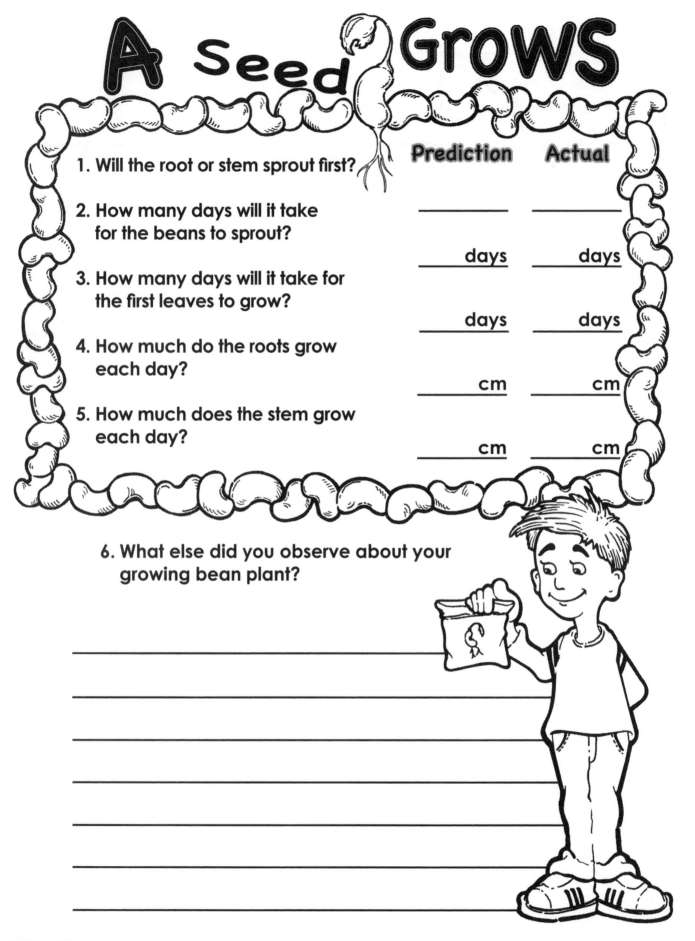

A Seed Grows

A Seed Grows

How much does your seed grow each day? Measure the length of the stem and the root daily. Color the graphs to match the lengths.

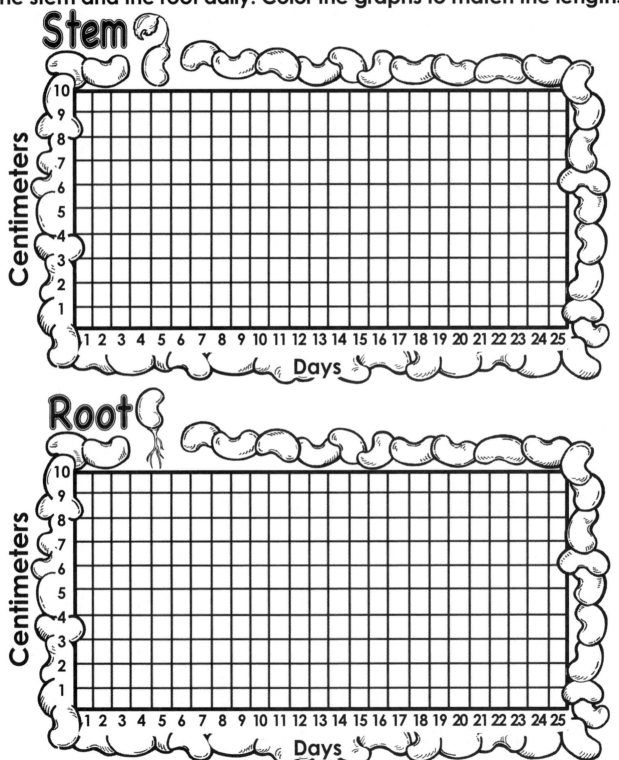

Connecting Learning

1. Describe the growth of the roots and the stem.

2. How do the roots and stems compare in color and direction?

3. How long did it take for your seed to sprout? Did everyone's seeds sprout on the same day? Why do you think there were differences?

4. How did you provide for the needs of the seed for growth?

5. What are you wondering now?

A Plant Begins

Topic
Plants

Key Question
How do my seeds grow?

Learning Goal
Students will observe and record the growth of seeds.

Guiding Documents
Project 2061 Benchmarks
- Describing things as accurately as possible is important in science because it enables people to compare their observations with those of others.
- Plants and animals both need to take in water, and animals need to take in food. In addition, plants need light.

NRC Standard
- Organisms have basic needs. For example, animals need air, water, and food; plants require air, water, nutrients, and light. Organisms can survive only in environments in which their needs can be met. The world has many different environments, and distinct environments support the life of different types of organisms.

Science
Life science
  botany
    plant growth

Integrated Processes
Observing
Comparing and contrasting
Communicating

Materials
Seeds (lima, corn, radish)
Potting soil
Daily log
Container (see *Management 1*)
Watering container or misting bottle

Background Information
Seeds start to grow when conditions are right to support the needs of growing plants. Water, air, and proper temperature are all necessary for seed growth. Germination rates vary according to type of seed used, the amount of water given, and the temperature.

Management
1. Each student should have a container in which to plant a seed. Styrofoam cups, plastic cups, and milk cartons all work well.
2. Prepare a daily log for each child. Duplicate several of the recording pages. Cut the papers in half and staple them inside the cover to make a logbook.
3. Misting bottles will help to prevent students from overwatering their plants.
4. By using a variety of different types of seeds, students can compare the germination rates. They can also compare what the various plants look like.

Procedure
1. Ask the *Key Question* and state the *Learning Goal*.
2. Distribute the materials and the page of instructions.
3. Direct the students to follow the directions to plant their seeds. Caution students against overwatering.
4. Have students assemble their Daily Logs.
5. Explain to students that when their seeds sprout and grow above the soil, they will start recording the growth by drawing and writing their observations.
6. Have students continue to record their observations every few days.

Connecting Learning
1. How many days did it take for your seeds to sprout?
2. Did they all sprout on the same day? Explain.
3. What did the sprouts look like?
4. Who else had plants like yours? How did you know they were the same?
5. How many different types of plants did we use?
6. Did we have any plants that didn't sprout? What might have caused that?
7. What are you wondering now?

A Plant Begins

Key Question

How do my seeds grow?

Learning Goal

Students will:

observe and record the growth of seeds.

A Plant Begins

You will need:

Seeds

Potting soil

Daily Log

Container

(lima, corn, or radish)

(plastic pot, Styrofoam cup, or milk carton)

Do this:

1. Fill the pot with potting soil.
2. Plant the seeds.
3. Keep the soil moist.
4. Make your daily log.
5. When the seeds sprout and grow above the soil, start recording the growth by drawing and writing your observations.
6. Continue recording observations every few days.

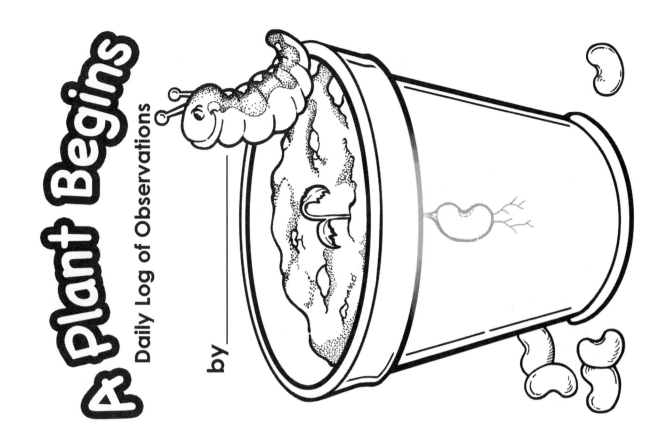

A Plant Begins
Daily Log of Observations

by _____

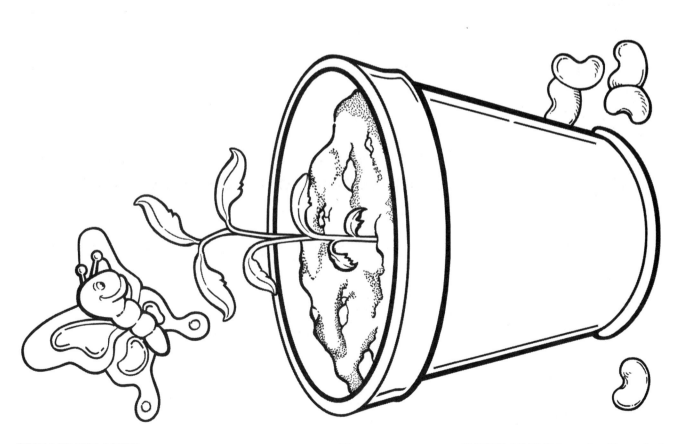

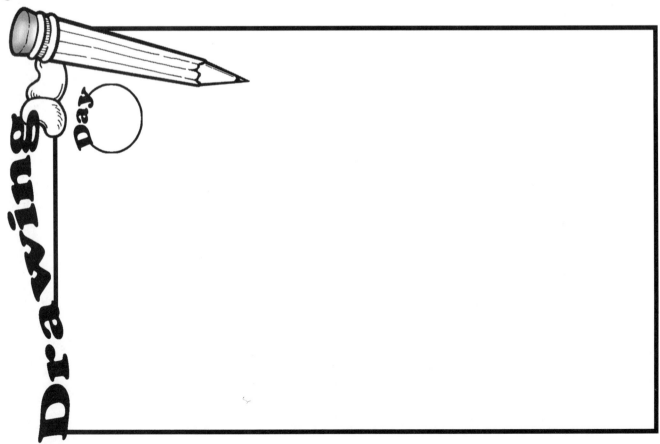

Connecting Learning

1. How many days did it take for your seeds to sprout?

2. Did they all sprout on the same day? Explain.

3. What did the sprouts look like?

4. Who else had plants like yours? How did you know they were the same?

5. How many different types of plants did we use?

6. Did we have any plants that didn't sprout? What might have caused that?

7. What are you wondering now?

It's in the Bag

Topic
Seed growth

Key Question
What does a seed look like as it starts to grow?

Learning Goals
Students will:
- "plant" seeds in a plastic bag; and
- observe and measure the growth of roots, stems, and leaves.

Guiding Documents
Project 2061 Benchmarks
- *Describing things as accurately as possible is important in science because it enables people to compare their observations with those of others.*
- *Plants and animals both need to take in water, and animals need to take in food. In addition, plants need light.*

NRC Standards
- *Employ simple equipment and tools to gather data and extend the senses.*
- *Organisms have basic needs. For example, animals need air, water, and food; plants require air, water, nutrients, and light. Organisms can survive only in environments in which their needs can be met. The world has many different environments, and distinct environments support the life of different types of organisms.*

NCTM Standards 2000*
- *Recognize the attributes of length, volume, weight, area, and time*
- *Recognize and apply mathematics in contexts outside of mathematics*
- *Use tools to measure*

Math
Measurement
  length

Science
Life science
  botany
    seed growth

Integrated Processes
Observing
Comparing and contrasting
Collecting and recording data
Interpreting data
Drawing conclusions

Materials
Assorted seeds (see *Management 1*)
Plastic zipper-type bags, pint size
Transparent tape
Paper towels
Seed ruler (see *Management 6*)
Paper clip

Background Information
The production of seeds is the last stage of reproduction of flowering plants. When fertilization takes place, the ovules of a flower become seeds. These are called embryos, which means new developing plants. Surrounding the embryo is a layer of cells called the endosperm where food is stored for the embryo to use later.

The outer layer of the ovule becomes the seed coat, the skin that provides protection for the seed. When seeds leave the parent plant, they become dormant until just the right conditions of warmth and water cause them to germinate.

Management
1. Use seeds such as lima beans, kidney beans, corn, sunflower seeds, garbanzo beans, etc. When selecting an assortment of seeds, use medium to large size seeds. The broad bean type are easier for young students to handle. At least two types of seeds are needed.
2. This activity is most successful if each student has a bag with seeds.
3. Placing seeds in a moist warm environment (in the bag) may encourage the growth of mold. Rinsing seeds in a very weak bleach solution (one teaspoon bleach to one gallon of water) will inhibit unwanted mold growth.
4. Make several copies of the page *How my Seeds Grow* so the students can staple the pages together to make a booklet.
5. Each student will need a 10-centimeter (4-inch) strip of transparent tape.
6. Copy the seed rulers onto card stock or transparency film. Each student will need one ruler.

PRIMARILY PLANTS © 2005 AIMS Education Foundation

Procedure
1. Ask the *Key Question*.
2. Show students the selection of seeds and invite students to choose two seeds that are different from each other.
3. Give each student a plastic bag and a paper towel and demonstrate how to fold the paper towel so it will fit inside the plastic bag.
4. Have students put their two seeds in the approximate middle of the paper towel. Distribute the transparent tape and direct students to place the tape across the seeds to adhere them to the paper towel.
5. Help students slip the paper towel with seeds into the bag. Have them dampen the paper towel with water. (Caution students about putting too much water in the bag. The towel should be damp without any excess water collecting in the bottom of the bag.) Tell them to zip the bag closed.
6. Invite students to tape the bags to the window or hang them on a line along the windows.
7. Distribute the measuring rulers. Have students attach the ruler to the plastic bag with a paper clip.
8. Allow a period for observation and measurement each day or every other day, depending on how rapid growth changes occur. Make sure that students realize that the 0 on the ruler is placed next to the seed. The numbers that go down are for measuring the roots while the numbers that go up are for measuring the stem.
9. Have students assemble the book *How my Seeds Grow*. Have students record their observations with pictures and measurements.
10. When plants outgrow the plastic bag, have students transplant them in pots or a terrarium.

Connecting Learning
1. Which seed sprouted first?
2. Which came first, the root or the stem?
3. Why do you think we didn't have to water the plants while they were in the plastic bags?
4. Plants need water, light, and food. How were their needs met while they were in the bag?
5. What are you wondering now?

* Reprinted with permission from *Principles and Standards for School Mathematics*, 2000 by the National Council of Teachers of Mathematics. All rights reserved.

It's in the Bag

Key Question

What does a seed look like as it starts to grow?

Learning Goals

Students will:

- "plant" seeds in a plastic bag; and
- observe and measure the growth of roots, stems, and leaves.

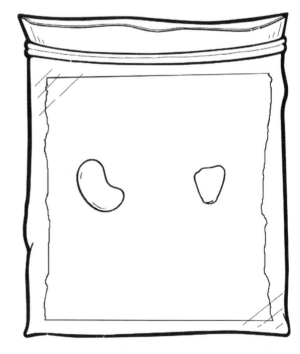

It's in the Bag
How My Seed Grows

Print extra copies of this page. Cut on the solid lines and staple to make a record book. After the seeds begin to sprout, have students record the growth by measuring and drawing every other day.

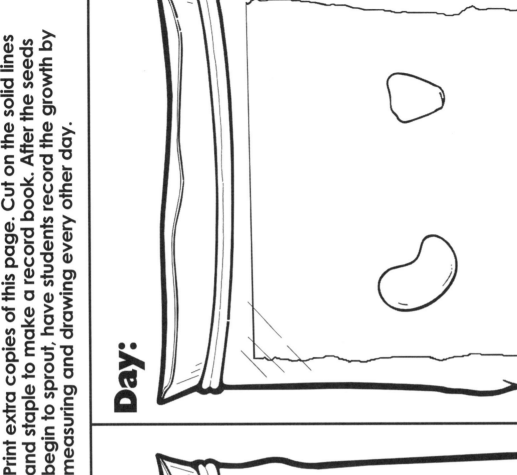

Day:

Day:

PRIMARILY PLANTS   27   © 2005 AIMS Education Foundation

It's in the Bag — How My Seed Grows

10	10	10	10	10	10	10	10
9	9	9	9	9	9	9	9
8	8	8	8	8	8	8	8
7	7	7	7	7	7	7	7
6	6	6	6	6	6	6	6
5	5	5	5	5	5	5	5
4	4	4	4	4	4	4	4
3	3	3	3	3	3	3	3
2	2	2	2	2	2	2	2
1	1	1	1	1	1	1	1
0	0	0	0	0	0	0	0
1	1	1	1	1	1	1	1
2	2	2	2	2	2	2	2
3	3	3	3	3	3	3	3
4	4	4	4	4	4	4	4
5	5	5	5	5	5	5	5
6	6	6	6	6	6	6	6
7	7	7	7	7	7	7	7
8	8	8	8	8	8	8	8
9	9	9	9	9	9	9	9
10	10	10	10	10	10	10	10

PRIMARILY PLANTS © 2005 AIMS Education Foundation

Connecting Learning

1. Which seed sprouted first?

2. Which came first, the root or the stem?

3. Why do you think we didn't have to water the plants while they were in the plastic bags?

4. Plants need water, light, and food. How were their needs met while they were in the bag?

5. What are you wondering now?

Little Brown Seeds

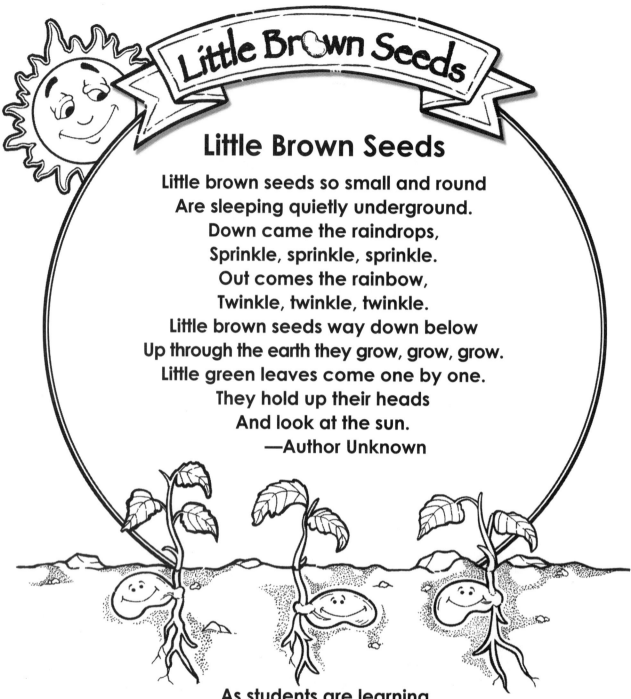

Little Brown Seeds

Little brown seeds so small and round
Are sleeping quietly underground.
Down came the raindrops,
Sprinkle, sprinkle, sprinkle.
Out comes the rainbow,
Twinkle, twinkle, twinkle.
Little brown seeds way down below
Up through the earth they grow, grow, grow.
Little green leaves come one by one.
They hold up their heads
And look at the sun.
—Author Unknown

As students are learning
about seeds and growing plants,
they can learn the poem *Little Brown Seeds*.

1. Print the pages of the booklet. Cut the pages in half and have students put them in sequential order.

2. Students can color the pictures or glue construction paper cut outs and the actual seeds and soil on the pictures.

PRIMARILY PLANTS   © 2005 AIMS Education Foundation

Little brown seeds so small and round
Are sleeping quietly underground.

1.

Down came the raindrops
Sprinkle, sprinkle, sprinkle.

2.

Out comes the rainbow,
Twinkle, twinkle, twinkle.

3.

Little brown seeds way down below
Up through the earth they grow, grow, grow.

4.

Little green leaves come one by one.

5.

They hold up their heads
And look at the sun.

6.

Soak and Sprout

Topic
Seed sprouting

Key Questions
Which of our seeds will sprout and which will not? Why?

Learning Goals
Students will:
- discover if soaked seeds and dry seeds will sprout, and
- determine if whole seeds and split seeds will sprout.

Guiding Documents
Project 2061 Benchmark
- *Things can be done to materials to change some of their properties, but not all materials respond the same way to what is done to them.*

NRC Standards
- *Employ simple equipment and tools to gather data and extend the senses.*
- *Use data to construct a reasonable explanation.*
- *Communicate investigations and explanations.*

Science
Life science
  seeds

Integrated Processes
Observing
Recording
Comparing and contrasting
Communicating

Materials
For each group of students:
  cups of soaked and dried seeds (see *Management 1*)
  Science Investigation Logs (see *Management 3*)
  hand lens
  plastic knife

For the class:
  two types of dried seeds (see *Management 1*)

Background Information
All seeds consist of three parts: the seed coat, the embryo, and the cotyledons (food storage). The seed coat protects the embryo inside the seed. Once the seed is exposed to water, the seed coat begins to soften and absorb the water. When the seed is in a warm, moist environment, germination occurs. The embryo begins to grow, breaking through the softened coat. First a root appears and then a sprout begins to emerge. The embryo feeds on the food stored in the cotyledons. It is necessary for the entire seed to be intact in order for this process to occur. Once the shoot is above the ground, the stem straightens and the plant's first leaves begin to grow and enable the plant to produce its own food. The cotyledons, sometimes called seed leaves, fall off the stem at this time.

In this activity, the students take a close look at the inside of a seed, locating the embryo. The students will discover through their observations that the split seeds do not contain the embryo which is vital to the sprouting of seeds. (A pea is a dicotyledon—a seed with two cotyledons. Each split pea is actually one-half of the pea; one cotyledon of the pea. When the seed is split apart, the embryo is damaged and falls off.) The lesson also gives the students an opportunity to observe the necessity of moisture in the process of germination. They will discover that soaked seeds will sprout and dry seeds will not sprout.

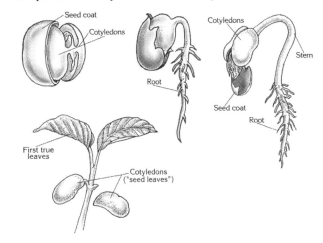

Management
1. A three-ounce cup of each of two types of seeds (dried black-eyed peas and dried split peas) will need to be soaked for one day prior to this activity. Each group will need two cups of soaked seeds and two cups of dried seeds.
2. If the class experiences the activity *Seed Soakers* prior to this lesson, review the observations they made.
3. For each group, duplicate a *Science Investigation Log*. Decide if you will be asking the students to make two, three, or more observations of the seeds. Duplicate one log for each scheduled observation. Direct the students to fill in the day in the box provided, (e.g., *Day 1*), and mark the day on the calendar appropriate for the month, (e.g., If they begin the investigation on September 23, they would mark this day on the calendar) and to record the date on the line provided.

Procedure

Part One
1. Ask the class to explain how they think a plant can grow out of the hard case of the seed.
2. To each group, distribute hand lenses and the small cups of dried black-eyed peas and dried split peas. Direct the students to use a hand lens and to closely examine the seeds. Discuss the hardness of the outside of the seeds. Record their observations on a large piece of chart paper in front of the class.
3. Tell the students they are going to conduct an investigation to observe what happens to the seeds after they have been soaked in water. Explain that each group of students will be using two different types of seeds and that they will need to observe their seeds very closely.
4. To each group, distribute the cups of soaked seeds. Explain to the students that these seeds have been soaking for a day. Invite the students to remove a soaked black-eyed pea and a split pea. Ask them to compare these seeds to the dry seeds. Discuss the hardness of the outside of the seed, the size, the color, and other observations.
5. Discuss whether or not they think each type—dried and soaked—of seed will be able to sprout and grow.

Part Two
1. Direct the students to drain off any water left in the cups of seeds and to store their cups of soaked seeds in a warm, dark area.
2. Instruct the students to also place a cup of each of the dried seeds next to the cups of soaked seeds.
3. The next day, using a plastic knife, demonstrate how to open one of the soaked black-eyed peas. Direct the students to open one or two of their soaked peas. Tell them to use a hand lens to help them examine the inside of the seeds. Ask the students to describe what they observe. Guide them to notice the tiny baby-like plant (embryo). Discuss the embryo, the seed coat, and the cotyledon (stored food). Explain the role each part plays in the germination process as described in the *Background Information* of this lesson.
4. Direct them to try to locate the embryo in their soaked split peas. Discuss how the pea is already split apart and that there are no embryos. Ask the students to explain what difference they think this will or will not make in the germination process.
5. Give each student an *Investigation Log*. Direct them to record Day 1, to mark the day on their calendar, and to record the date on the line provided. Remind them to also list the members of their science team on the cover. Direct their attention to the inside pages of their *Investigation Logs*. Tell them to record whether or not any of their seeds have sprouted and to write about their observations, their thinking and any conclusions they can make at this time. Suggest they write about their observations of the inside of the seeds.
6. After a few days, ask the students to observe their seeds. Discuss the changes that have occurred. Have them look for sprouting seeds, roots, etc. Discuss the results, and direct the students to use a second copy of the *Science Investigation Log* to record their findings.
7. Continue for several days, choosing at least one other day to record observations.
8. Discuss the results of the investigation.

Connecting Learning
1. What did you think would happen in the investigation? Did you think both types of seeds would sprout or would only one or none of them sprout? Why did you think this?
2. Describe what you observed about the seeds in your investigation.
3. What happened to the soaked seeds compared to the seeds that are still dry?
4. Why do you think the dried seeds have a hard coat (outside)? [to protect the embryo inside the seed]
5. Describe the parts of the seed and how these parts work together in the germination process.
6. How did the hand lens help you look at the seeds?
7. Did all the seeds sprout? Why do you think this happened?
8. What do you think will happen if you leave the seeds alone for several more days?
9. What do you think happens to seeds once they are planted in the ground?
10. Why do you think some seeds planted in the ground sprout and some don't?
11. What do you think we should do with the seeds now?

Curriculum Correlation

Ganeri, Anita. *What's Inside Plants?* Peter Bedrick Books. New York. 1993. Pictorial and technical text descriptions of the anatomy of many different plants.

Kuchalla, Susan. *Now I Know All About Seeds.* Troll Associates. New York. 1982. Simple text and child-centered pictures present several kinds of seeds and show how they grow into plants.

Pascoe, Elaine. *Seeds and Seedlings.* Blackbirch Press, Inc. Woodbridge, CT. 1997. Describes how seeds are formed, how they grow, what they look like, how they reproduce, and how they make food.

Soak and Sprout

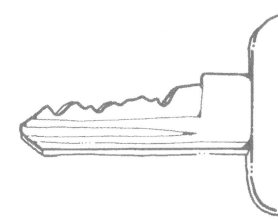

Key Questions

Which of our seeds will sprout and which will not? Why?

Learning Goals

Students will:

- discover if soaked seeds and dry seeds will sprout, and
- determine if whole seeds and split seeds will sprout.

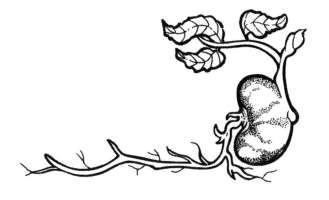

Soak and Sprout
Science Investigation Log

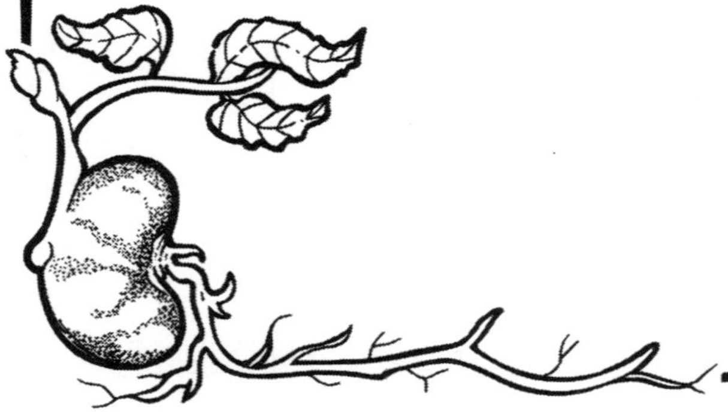

Science team members:

Day

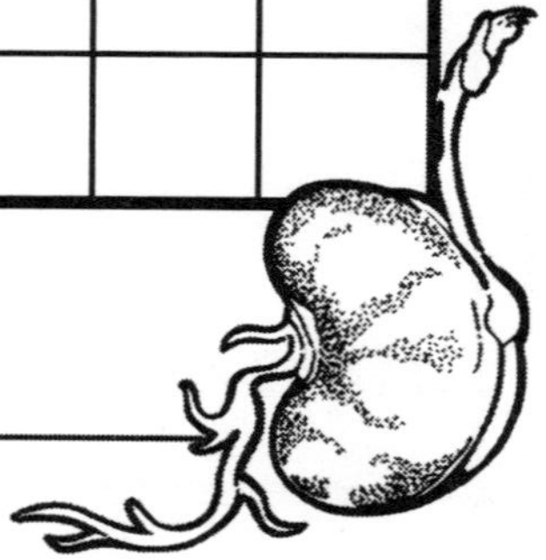

Date _____

Our thinking... _____

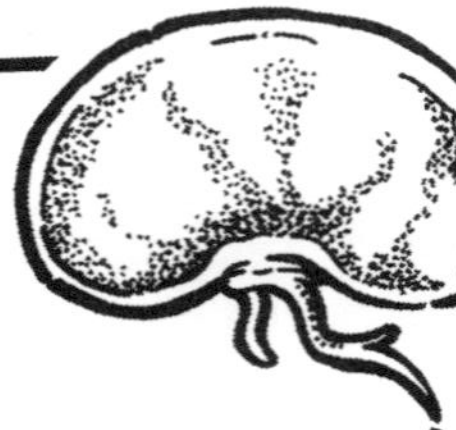

Our conclusions... _____

PRIMARILY PLANTS

Write your observations.

Did the seeds sprout?

Soaked Seeds
yes   no
Whole seeds
Split seeds

Dried Seeds
yes   no
Whole seeds
Split seeds

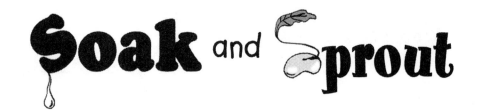

Connecting Learning

1. What did you think would happen in the investigation? Did you think both types of seeds would sprout or would only one or none of them sprout? Why did you think this?

2. Describe what you observed about the seeds in your investigation.

3. What happened to the soaked seeds compared to the seeds that are still dry?

4. Why do you think the dried seeds have a hard coat (outside)?

5. Describe the parts of the seed and how these parts work together in the germination process.

Connecting Learning

6. How did the hand lens help you look at the seeds?

7. Did all the seeds sprout? Why do you think this happened?

8. What do you think will happen if you leave the seeds alone for several more days?

9. What do you think happens to seeds once they are planted in the ground?

10. Why do you think some seeds planted in the ground sprout and some don't?

11. What do you think we should do with the seeds now?

Seeds, Spores, and More.

A seed is a tiny case containing plant life. It contains a small plant and food. Seeds come in all different sizes and shapes, but when you open a seed case, you will find that every seed contains a tiny plant called an embryo. The seeds contain food to sustain the tiny plant until it can make its own. The food storage is called the cotyledon. Some plants, such as a bean, have two cotyledons. Others, such as corn, have only one cotyledon.

Nearly all trees, shrubs, vegetables, and flowers started as seeds. Some of the seeds grow into seedlings and then into adult plants. Very few seeds actually survive, so plants produce an enormous number of seeds to ensure reproduction of the plant.

When a seed is ripe, it drops from the parent plant. It starts to grow or germinate when it has water, warmth, and air. A root appears first and grows downward. Then a stem pushes upward toward the light. The first leaves appear and the plant can make its own food in its leaves.

There are many kinds of seeds that we eat. Perhaps the most important plants in the world are grasses. Wheat, oats, corn, and rice are different kinds of grasses called cereals. The seeds of these cereals provide food for many of the animals we eat.

There are also non-flowering plants such as ferns, mosses, and algae that produce spores. A spore is a cell with a thick, protective covering, much smaller than the smallest seed. Spores do not include an embryo, so they develop directly into an adult plant. When conditions are right, spores are distributed by air or water.

On the underside of fern leaves, you can find small brown lumps. These lumps produce spores. When the spores are ripe, they fall to the ground and grown into new ferns.

Seed Sort

Topic
Seeds

Key Question
How are seeds alike and different?

Learning Goals
Students will:
- count and sort seeds, and
- find likenesses and differences of many seeds.

Guiding Documents
Project 2061 Benchmarks
- *Some animals and plants are alike in the way they look and in the things they do, and others are very different from one another.*
- *Numbers can be used to count things, place them in order, or name them.*
- *Simple graphs can help to tell about observations.*

NCTM Standards 2000*
- *Count with understanding and recognize "how many" in sets of objects*
- *Sort and classify objects according to their attributes and organize data about the objects*

Math
Measurement
Counting
Estimation

Science
Life science
  botany
    diversity

Integrated Processes
Observing
Predicting
Sorting and classifying
Collecting and recording data
Comparing and contrasting

Materials
Small paper cups, 3-oz
Assorted seeds (see *Management 1*)
Hand lenses
Balance
Teddy Bear Counters

Background Information
Seeds are all different sizes and shapes and they come surrounded by all different kinds of fruit. But all seeds are alike in two ways. Every seed contains a little plant called an embryo, and all seeds contain food that helps the little plant grow.

Seeds are remarkable in the ways in which they spread themselves in order to grow new plants. Some seeds simply fall to the ground, others float on water, some are fired like buckshot over a distance, and others attach to an animal's fur.

All seeds serve the same purpose—to germinate and grow a new plant in order to perpetuate the plant species.

Management
1. Have at least five or six different kinds of seeds. The student pages use lima beans, kidney beans, popcorn, sunflower seeds, garbanzo beans, black-eyed peas.
2. Mix seeds together so students each get a variety of types. Fill the small paper cups about two-thirds full.
3. Place students in groups of three to four.

Procedure
1. Distribute the cupfuls of seeds.
2. Have the students estimate how many seeds are in the cup.
3. Give each group a copy of the first student page.
4. Tell students to pour the cup of seeds into the flower in the middle of the paper and sort the seeds into smaller sets of like kinds and put them in the smaller circles.
5. Have them count the number of seeds in each small circle.
6. Direct the students to add all the seed groups together and record the total count of seeds that were in the cup.

PRIMARILY PLANTS © 2005 AIMS Education Foundation

7. Invite the students to use a balance to find out how many seeds it takes to equal the mass of a Teddy Bear Counter. Have them record their estimations and findings on the second student page.
8. Distribute the hand lenses and third student page. Have students look closely at the seeds and record their colors and shapes.
9. Tell students that they will now see how many seeds it takes to cover a line. Encourage them to first predict and then find the actual.

Connecting Learning
1. Of which seed type were there the most in your group? …the least?
2. How are the seeds alike? How are they different?
3. What colors are the seeds?
4. Are the seeds all the same shape? Explain.
5. Are they all the same size? Explain.
6. What are you wondering now?

* Reprinted with permission from *Principles and Standards for School Mathematics*, 2000 by the National Council of Teachers of Mathematics. All rights reserved.

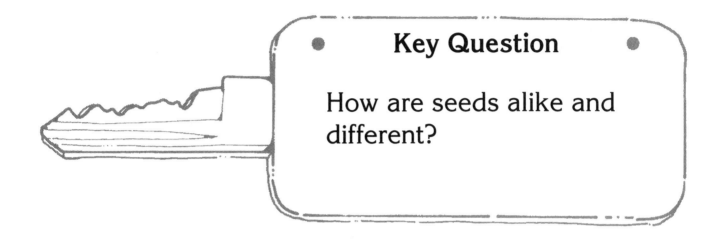

Key Question

How are seeds alike and different?

Learning Goals

Students will:

- count and sort seeds, and
- find likenesses and differences of many seeds.

Seed Sort

1. How many seeds do you think are in your cup?

We think there are _____ seeds.

2. Sort your seeds into groups. Count each group.

3. Add all the seed groups together.

We count _____ seeds in all.

Put all your seeds here. Then sort them into the smaller circles.

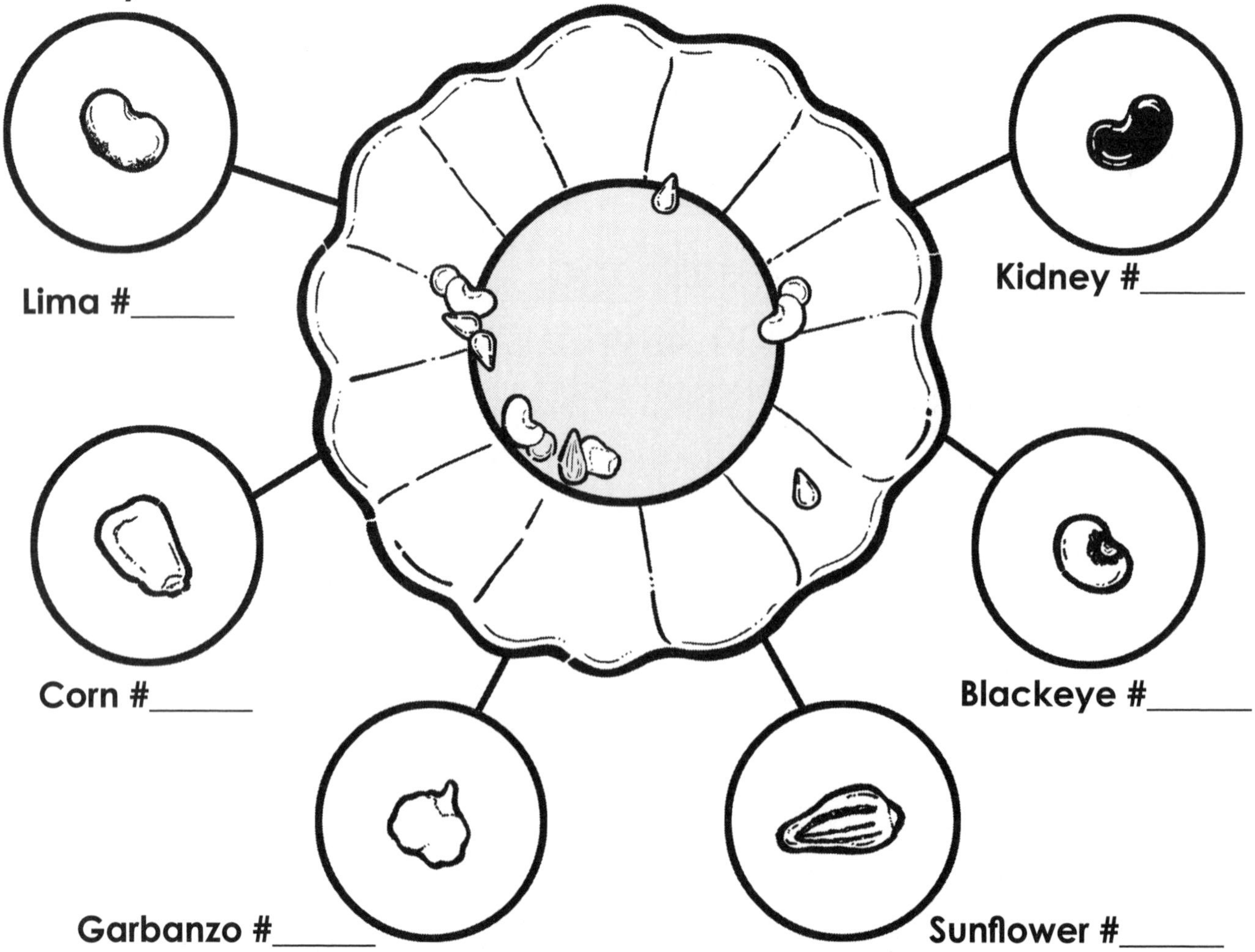

Lima #_____

Kidney #_____

Corn #_____

Blackeye #_____

Garbanzo #_____

Sunflower #_____

PRIMARILY PLANTS © 2005 AIMS Education Foundation

Seed Sort
Comparing Seeds

How many of each seed will balance one Teddy Bear Counter?

	I think:	I count:
Lima	◯	◯
Garbanzo	◯	◯
Corn	◯	◯
Sunflower	◯	◯
Blackeye	◯	◯
Kidney	◯	◯

Color the graph to show what you found out.
Number of seeds to balance one Teddy Bear Counter

	1	2	3	4	5	6	7	8	9	10	11	12	13	14	15	16	17	18	19	20
Lima																				
Garbanzo																				
Corn																				
Sunflower																				
Blackeye																				
Kidney																				

PRIMARILY PLANTS © 2005 AIMS Education Foundation

Seed Sort

Observing Seeds

Look closely at your seeds. Name the colors that you see. Draw the shapes of the seeds.

	Colors	Shapes
Lima		
Garbanzo		
Corn		
Sunflower		
Blackeye		
Kidney		

How many seeds does it take to cover the line below? _____

	I think:	I count:
Lima		
Garbanzo		
Corn		
Sunflower		
Blackeye		
Kidney		

PRIMARILY PLANTS © 2005 AIMS Education Foundation

Connecting Learning

1. Of which seed type were there the most in your group? ...the least?

2. How are the seeds alike? How are they different?

3. What color are the seeds?

4. Are the seeds all the same shape? Explain.

5. Are they all the same size? Explain.

6. What are you wondering now?

The Seed Within

Topic
Seeds

Key Question
How do the seeds vary in size, color, and number in each fruit or vegetable?

Learning Goal
Students will compare the size, shape, and color of various seeds.

Guiding Documents
Project 2061 Benchmarks
- *Numbers can be used to count things, place them in order, or name them.*
- *Simple graphs can help to tell about observations.*

NRC Standard
- *Each plant or animal has different structures that serve different functions in growth, survival, and reproduction. For example, humans have distinct body structures for walking, holding, seeing, and talking.*

NCTM Standards 2000*
- *Count with understanding and recognize "how many" in sets of objects*
- *Represent data using concrete objects, pictures, and graphs*

Math
Counting
Graphing

Science
Life science
   botany
      seeds

Integrated Processes
Observing
Collecting and recording data
Interpreting data
Comparing and contrasting

Materials
Fresh fruits and vegetables (see *Management 1*)
Sharp knife
Paper towels
Paper plates

Background Information
    Seeds are made in the fertilized ovule of a flower. Once a flower has been pollinated, seeds begin to develop. The part of the flower that holds the seeds starts to grow bigger. This part becomes the fruit, the protective structure surrounding the seeds. Plant species survive because of the seeds the fruit protects.
    The fruits can be fleshy and moist like apples, melons, and tomatoes, or it can be hard and dry like nuts and beans. Animals readily eat the sweet, juicy, and brightly colored fruits and so the seeds are scattered far from the parent plant. Although we use the terms fruits and vegetables for the protective structure surrounding the seeds, in reality all of them are the fruit of the plant.
    The size of the seed does not necessarily indicate the size of the parent plant. A peach seed produces a peach tree. Yet, a coastal redwood seed is much smaller and produces some of the largest trees on the surface of the Earth. The size of the seed does not necessarily indicate the size of the fruit. The peach seed is relatively large compared to its fruit. The watermelon seed is relatively small compared to its fruit.

Management
1. There are two versions of the first student page provided. The first version has the fruits and vegetables drawn in, and the second is blank. Use the page that is most suited to your needs.
2. If you use the version with the picture drawn in, each group will need half of a tomato, bell pepper, apple, and orange, a slice of melon, and a whole peach, avocado, and pea pod (these can be shared by groups).
3. Divide students into groups; part of the group can remove the seeds, the other part can count the seeds.

PRIMARILY PLANTS   © 2005 AIMS Education Foundation

4. Prepare a large class graph on which students can glue their seeds.
5. For young children, explore just one fruit a day. The fruit outlines can be enlarged to page size or to a large outline on the wall. The children can color in the seeds on the outlines. They can record the number of seeds in each fruit by coloring the graph on the activity page. They can glue their seeds to the large class graph placed on the wall or floor.

Procedure
1. Hold an introductory discussion:
   a. How do plants start?
   b. Where do we get the seeds?
   c. Do all fruits and vegetables have the same number of seeds?
   d. Why do some fruits and vegetables have many seeds and others have just a few?
   e. Are all seeds the same size?
   f. In what ways are seeds the same or different?
2. Show the students the various fruits and vegetables. List them on the board or on a transparency.
3. Ask the students what they think the insides look like.
4. Cut the fruits and vegetables and distribute the pieces to the students for observation. Allow time for the students to observe their fruits.
5. Tell students to record where the seeds are located on their student pages.
6. Have them count the seeds and record the number and shape on the page.
7. Record the class results on a wall chart or transparency.
8. Wash and dry the seeds. Invite the students to glue the seeds to the large class graph.

Connecting Learning
1. Where were the seeds located?
2. How were the seeds different?
3. What shape are the sees?
4. How many seeds does each fruit or vegetable have?
5. Do all large fruits and vegetables have large seeds? What about small fruits and vegetables?
6. Which fruits have many seeds in them?
7. Which fruits have only one seed in them?
8. Can you think of a large fruit or vegetable that produces a small seed? What is it?
9. Is there a small fruit that has a large seed?

Extensions
1. Let students try growing the seeds they remove from the fruits and vegetables.
2. Have the students make a fruit salad with the fruit they have been exploring.
3. Have sequence the seeds from largest to smallest.

* Reprinted with permission from *Principles and Standards for School Mathematics*, 2000 by the National Council of Teachers of Mathematics. All rights reserved.

The Seed Within

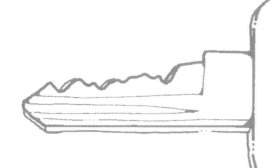

Key Question

How do the seeds vary in size, color, and number in each fruit or vegetable?

Learning Goal

compare the size, shape, and color of various seeds.

The Seed Within

1. Look closely at the fruits and vegetables.
2. Draw where you find the seeds.
3. Carefully count the seeds and record.
4. Wash and dry the seeds.

Avocado — # of seeds / Shape of seeds

Peach — # of seeds / Shape of seeds

Bell Pepper — # of seeds / Shape of seeds

Tomato — # of seeds / Shape of seeds

Peas — # of seeds / Shape of seeds

Orange — # of seeds / Shape of seeds

Apple — # of seeds / Shape of seeds

Melon — # of seeds / Shape of seeds

The Seed within

1. Look closely at the fruits and vegetables.
2. Draw where you find the seeds.
3. Carefully count the seeds and record.
4. Wash and dry the seeds.

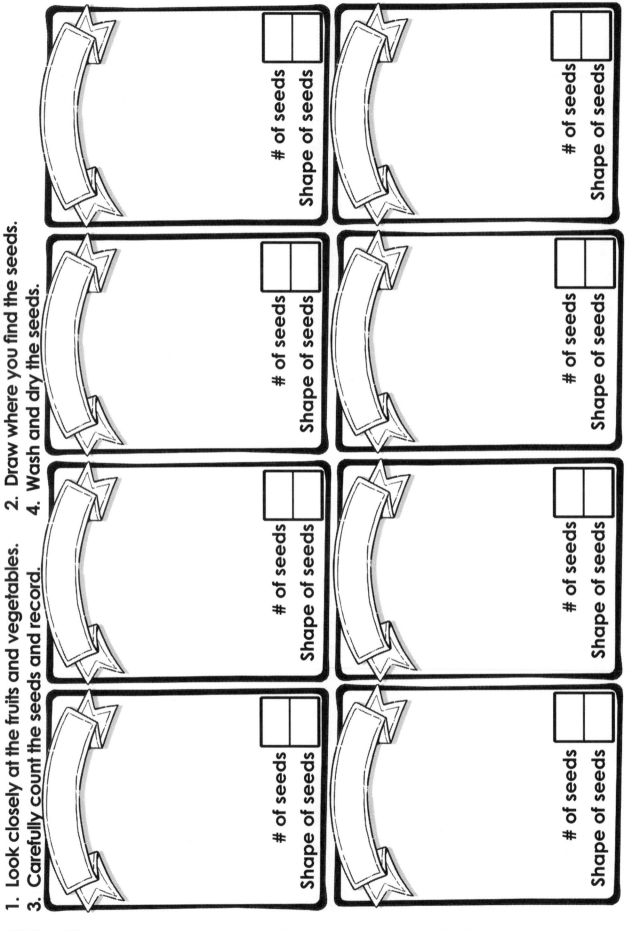

The Seed Within

Color your graph to show the number of seeds in each.

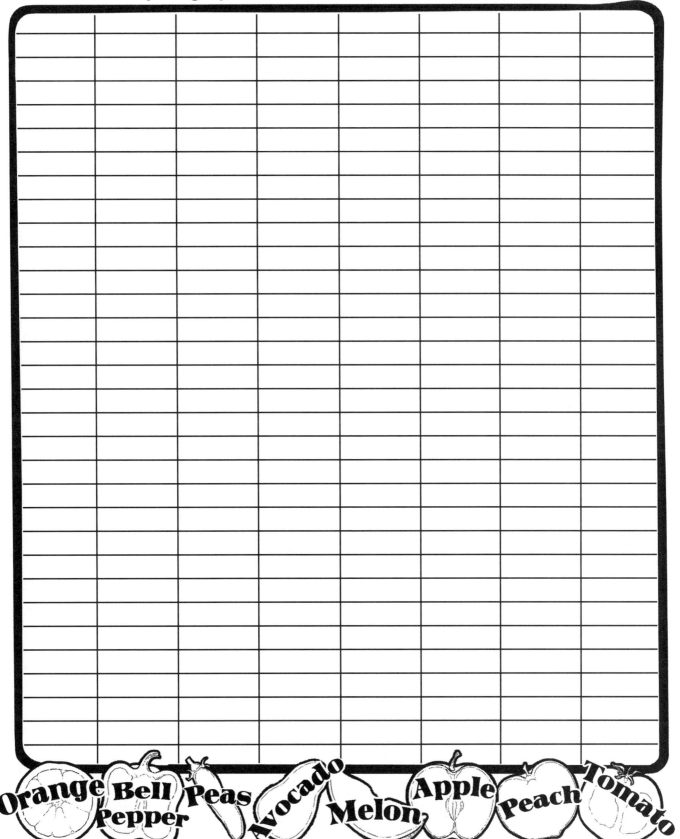

Orange, Bell Pepper, Peas, Avocado, Melon, Apple, Peach, Tomato

The Seed Within

Connecting Learning

1. Where were the seeds located?
2. How were the seeds different?
3. What shape are the sees?
4. How many seeds does each fruit or vegetable have?
5. Do all large fruits and vegetables have large seeds? What about small fruits and vegetables?
6. Which fruits have many seeds in them?
7. Which fruits have only one seed in them?
8. Can you think of a large fruit or vegetable that produces a small seed? What is it?
9. Is there a small fruit that has a large seed?

Seed Soakers

Topic
Seeds

Key Question
What happens to seeds when soaked in water?

Learning Goal
Students will observe changes that occur to seeds when soaked in water.

Guiding Documents
Project 2061 Benchmark
- Things can be done to materials to change some of their properties, but not all materials respond the same way to what is done to them.

NRC Standards
- Employ simple equipment and tools to gather data and extend the senses.
- Use data to construct a reasonable explanation.
- Communicate investigations and explanations.

NCTM Standards 2000
- Use tools to measure
- Recognize the attributes of length, weight, area, and time

Math
Measurement
  time

Science
Life science
  seeds

Integrated Processes
Observing
Recording
Comparing and contrasting
Communicating

Materials
For each group of students:
  two 3-ounce cups (see *Management 1*)
  Science Investigation Cards
  Science Investigation Log
  hand lens
  recloseable plastic bag, pint-size

For the class:
  dried seeds, black-eyed peas and split peas
    (see *Management 3*)
  chart paper
  transparent tape
  scissors
  a clock

Background Information
A seed contains the beginning of a plant and stored food to feed it. When a seed is ripe, it is released from the parent plant. It usually falls to the ground where it becomes embedded in soil. As the seed absorbs water, its coat softens and breaks apart to allow a sprout to emerge. This is called germination.

As the students investigate with different seeds, they will be able to make many observations, such as:
- different types of seeds absorb different amounts of water at different rates,
- the seed coat softens, and
- a change in the physical properties of the seeds occurs as they become swollen from the absorption of the water.

Management
1. Provide the same size and shape cups for each group. Three-ounce clear plastic cups can be purchased at hobby and grocery stores and provide an opportunity to see the entire contents of the cup of seeds.
2. For each group, duplicate and cut out one set of the *Science Investigation Cards* and a *Science Investigation Log*. Demonstrate how to assemble the log as illustrated. You will need to decide how often the class will be recording their observations (every 15 minutes or every 30 minutes or every hour, etc.) Each group will need one strip for each observation period. A total of three or four half-hour observations is suggested.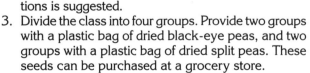
3. Divide the class into four groups. Provide two groups with a plastic bag of dried black-eye peas, and two groups with a plastic bag of dried split peas. These seeds can be purchased at a grocery store.

PRIMARILY PLANTS   57   © 2005 AIMS Education Foundation

Procedure
Part One
1. Using a sample of the two types of seeds, give one seed to each student.
2. Have the students examine the seeds and discuss with a partner the similarities and differences between their seeds. Give the students hand lenses to help them take a close look at the seeds. Discuss the hardness of the outside of the seeds. Record their observations on a large piece of chart paper in front of the class.
3. Ask the class to describe what a seed needs in order to grow. [water, soil, sun] Ask them to explain how they think a seed can grow a plant out of the hard case of the seed. Record their explanations on the class chart paper.
4. Tell the students they are going to conduct an experiment to observe what happens to the seeds when they are soaked in water. Explain to the class that they will be using two different types of seeds and that they will need to observe their seeds very closely.
5. Give each group a bag of one kind of seed, a set of *Investigation Cards,* and one *Investigation Log.*
6. Instruct each group to lay their *Investigation Cards* on the table in numerical order, to gather the materials they will need, and to follow the directions to begin the investigation. Explain that each group will be filling two cups with the same type of seeds, and that they will add water to only one of these cups. The cup of dried seeds will be used to compare to the cup of soaked seeds.
7. Tell them to start the recording of their investigation on the *Science Investigation Log* form labeled *1 Start.*
8. Once the investigations have been set up, the starting times have been recorded in their *Science Investigation Logs,* and illustrations have been made of the dried seeds, ask each group to detscribe the set-up of the investigation and to discuss the expected results.
9. At three or four 30-minute intervals, ask the students to observe their cup of seeds and to record the results and time of observation.
10. At the end of 60 minutes, direct the students to carefully add more water to any of the cups that have absorbed all the water that was first put in the cups of seeds. Encourage them to once again fill those cups to the top.
11. Direct the students to continue observing and recording their observations for another set of two 30-minute intervals. Discuss the results.
12. At the end of the investigation, ask the students to compare the cup of dried seeds to the cup of soaked seeds. Discuss what they observe. Remind them that the cup of dried seeds is the same amount of seeds they began with before soaking the seeds in the other cup.

Part Two
1. Direct each student to get a soaked seed from their cup. Ask them to carefully observe these seeds and to discuss the changes. Give the students hand lenses to help them take a close look at the seeds. Using the *Science Investigation Log* form labeled *End,* have them trace around one of the soaked seeds. Invite them to illustrate a detailed picture of this seed, paying attention to color and markings.
2. Discuss the hardness of the outside of the seed, the size, the color, etc. Have each student get a dried seed from their bag. Guide them to compare the dried seeds to the soaked seeds. Discuss the students' observations and have them illustrate the seeds in their *Science Investigation Logs.*

Connecting Learning
1. What did you do to the seeds? [We soaked them in water.]
2. How long did you soak the seeds before you noticed a change in the seeds?
3. How does your *Science Investigation Log* help you to think about this investigation? What kind of information did you record in the log?
4. What time was it when you started the investigation? What time was it when you ended the investigation?
5. Describe what you observed about the seeds in your investigation.
6. Why do you think the seeds changed?
7. Why do you think the cup of soaked seeds looks like there is more seeds than in the cup of dried seeds? Did you add any seeds to this cup? Do you think you grew more seeds? How could you find out?
8. What do you think happens to seeds once they are planted in the ground and watered?
9. How do the seeds that are planted outdoors get watered? [Rain or we water them with a hose, etc.]

Curriculum Correlation
Kuchalla, Susan. *Now I Know All About Seeds.* Troll Associates. New York. 1982. [Simple text and child centered pictures present several kinds of seeds and show how they grow into plants.]

Pascoe, Elaine. *Seeds and Seedlings.* Blackbirch Press, Inc. Woodbridge, CT. 1997. [Describes how seeds are formed, how they grow, what they look like, how they reproduce, and how they make food.]

* Reprinted with permission from *Principles and Standards for School Mathematics,* 2000 by the National Council of Teachers of Mathematics. All rights reserved.

Seed Soakers

Key Question

What happens to seeds when soaked in water?

Learning Goal

Students will:

observe changes that occur to seeds when soaked in water.

Fill 2 cups with seeds.

Fill 1 cup of seeds with water to the top.

Record in Science Log.

3

4

Observe both cups of seeds and record each time.

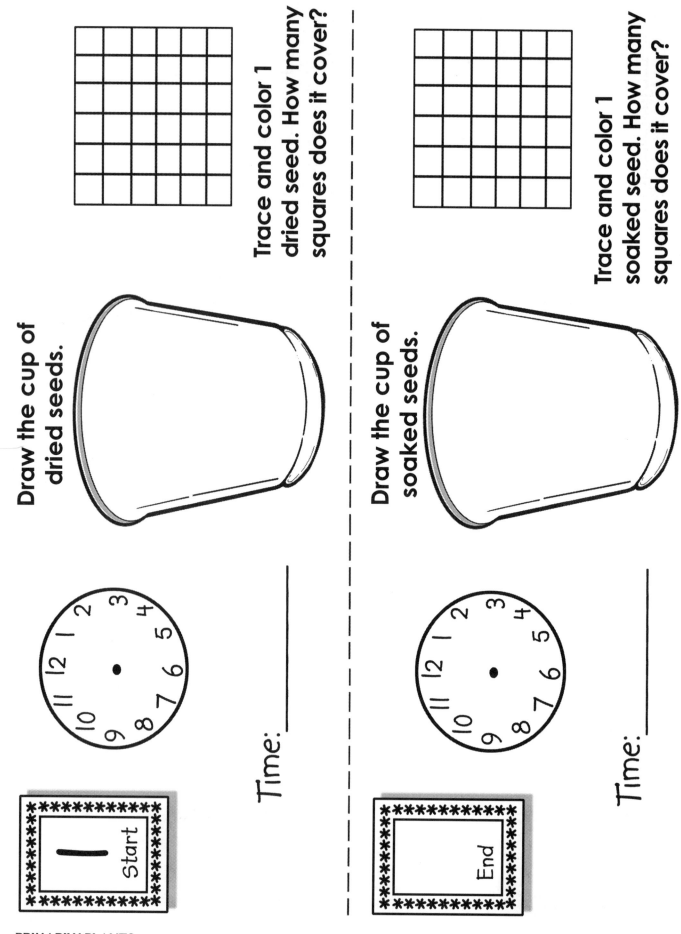

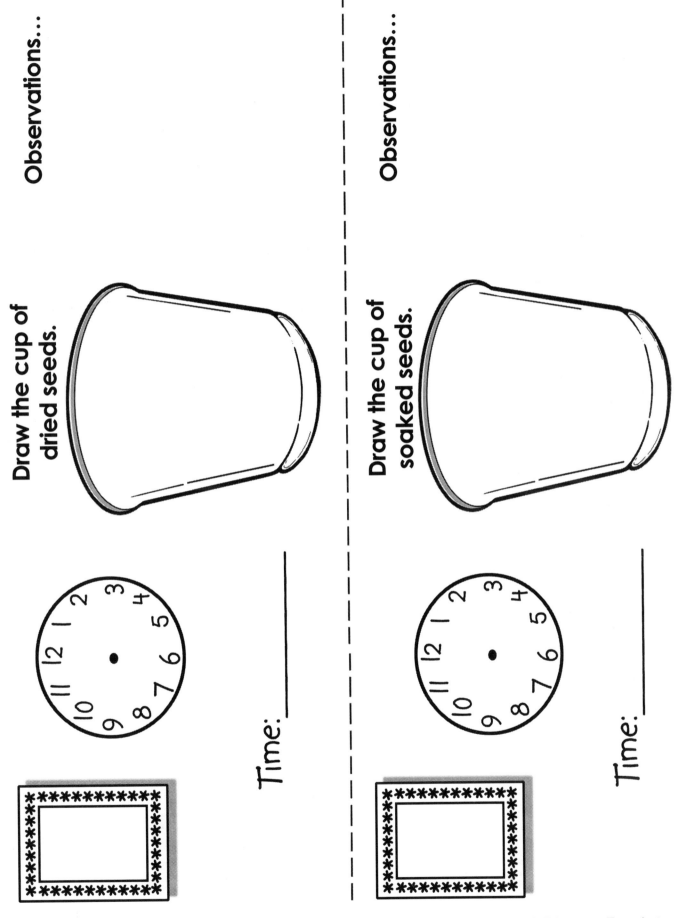

Connecting Learning

1. What did you do to the seeds?

2. How long did you soak the seeds before you noticed a change in the seeds?

3. How does your Science Investigation Log help you to think about this investigation? What kind of information did you record in the log?

4. What time was it when you started the investigation? What time was it when you ended the investigation?

5. Describe what you observed about the seeds in your investigation.

6. Why do you think the seeds changed?

Connecting Learning

7. Why do you think the cup of soaked seeds looks like there is more seeds than in the cup of dried seeds? Did you add any seeds to this cup? Do you think you grew more seeds? How could you find out?

8. What do you think happens to seeds once they are planted in the ground and watered?

9. How do the seeds that are planted outdoors get watered?

Seeds Travel

Topic
Seed dispersal

Key Question
How many ways can seeds be dispersed by a parent plant?

Learning Goal
Students will observe many ways that seeds travel from the parent plant.

Guiding Documents
Project 2061 Benchmarks
- Some animals and plants are alike in the way they look and in the things they do, and others are very different from one another.
- Plants and animals have features that help them live in different environments.

NRC Standard
- Each plant or animal has different structures that serve different functions in growth, survival, and reproduction. For example, humans have distinct body structures for walking, holding, seeing, and talking.

NCTM Standards 2000
- Use tools to measure
- Understand how to measure using nonstandard and standard units

Math
Measurement
 length
 time

Science
Life science
 botany
  seed dispersal

Integrated Processes
Observing
Comparing and contrasting
Recording data

Materials
Assorted seeds that travel by wind, water, animal fur (see *Management 1*)
Hand lenses
Metric measuring tape
Stopwatches
Student pages

Background Information
Most plants produce a large number of seeds. This is because so few seeds survive. In order to ensure survival, many seeds are modified in various ways so they can be carried away from their parent plant.

Some fruit and seeds simply drop from a parent plant. They take root there, but compete for space and light.

Many seeds have developed wings or silky hairs that allow them to be carried by the winds for miles. The dandelion seed, for example, has a little parachute that helps it to be carried by the wind.

Plants that grow along the banks of streams and rivers often have seeds that will float on water. The seeds usually have tough husks and air spaces to help them float. The best known seed that floats many miles is the coconut.

Many seeds have sharp hooks, or barbs, that stick to animals with furry coats, like sheep or dogs. They drop off some distance from where they grew.

Seed dispersal helps to prevent too many seedlings from growing in a small area near the parent plant. Those plant species that are able to spread their seeds widely have a better chance of surviving.

Management
1. This lesson will use only three of the many ways that seeds are dispersed. These three are ones that students will enjoy exploring and can easily observe.
2. Collect seeds from the different ways of seed travel: wind—dandelion, milkweed, maple, sycamore, pine; water—cranberry, coconut; animal fur—cocklebur, crabgrass, beggar-ticks, thistle.
3. This lesson needs to be taught in the fall season when many seeds can be collected, or the seeds can be collected to use later in the year.
4. To find seeds, look around your school. Check out the trees planted along the streets near the school, the weeds that survive along the edges of the playground, or the wild environment in a vacant lot.

PRIMARILY PLANTS

Procedure

Part One
1. Read the information sheet *Seeds Travel* with the students.
2. Display the seeds that are carried by animals' fur. Have the students use hand lenses to look at the tiny hooks or barbs on the seed pods. Discuss how these can hook into an animal's fur. Make sure they notice that each hooked bract has a seed at the bottom.
3. Ask students if they have ever walked across a lot or field that has tall weeds and grasses. Question them about whether they got some of these types of seeds in their socks.
4. Display the seeds that produce a "parachute." Have the students use hand lenses to look at the seed at the bottom of the parachute.
5. Encourage the students to let the seeds blow in the wind to see how far they will go before landing.
6. Display seeds that have "wings." If your school has a second story, let students drop the seeds from an upper window or stairwell. Dropping seeds from a set of bleachers also works well. If these are not available, invite the students to drop them from atop their chairs or tables. Caution students about safety issues. Guide students to observe that the seeds are like helicopters in the way they spin to the ground.
7. Display the seeds that float. Have the students observe the waxy waterproof coat of the cranberry. Cut it open so that students can see the four air pockets, each of which contains a seed.
8. Have students record their observations on the student sheets.

Part Two
1. Invite students to take two seeds that can travel by wind. Have them illustrate them on the student sheet *Helicopters and Parachutes*.
2. Distribute stopwatches. Direct students to perform the test to determine which seed can stay in the air longer.
3. Distribute metric measuring tapes. Have students go outside, drop both seeds, and measure how far they travel.

Connecting Learning
1. What are some ways that seeds travel?
2. Why do they need to travel?
3. How far did your winged seeds travel? Do they need a strong wind?
4. How do the seed cases (the fruit) differ with the various traveling methods?
5. Which do you think is the most efficient mode of travel for the seeds?
6. What is the best kind of weather for airborne seeds to disperse?
7. Can you find some way in which seeds are dispersed other than the ways that have been studied?
8. What are you wondering now?

Extension
1. Have the students collect as many kinds of seed as possible. Classify them into various ways of travel.
2. Write a story about a seed that travels far from the parent plant.
3. Make collages with the seeds that have been collected.
4. Make paper helicopters. Do they fly the same way that nature's seed helicopters do?
5. Choose a flowering plant in the spring. Watch it carefully. What happens when the petals fade? Where does the fruit form? What happens to the seeds? Go back in the fall. How are the seeds scattered—by the wind, by animals, or by water?

* Reprinted with permission from *Principles and Standards for School Mathematics*, 2000 by the National Council of Teachers of Mathematics. All rights reserved.

Seeds Travel

Key Question

How many ways can seeds be dispersed by a parent plant?

Learning Goal

observe many ways that seeds travel from the parent plant.

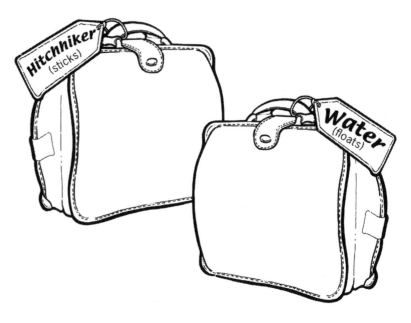

Seeds Travel

Seeds cannot move by themselves. They must be carried away from the parent plant so they have enough light and space to grow.

Air

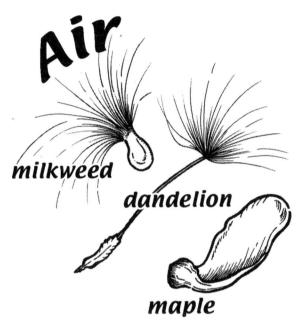

milkweed
dandelion
maple

Many light seeds have wings or silky hairs that help them to be carried by the wind. The hairs catch the wind like a parachute. The ones with wings turn like a helicopter as the seeds ride the wind away from the parent plant.

Water

cranberry
coconut

Some plants that live near water have seeds that float. The seeds drop into the water and float away from the parent plant. Some have spaces inside to help them float. Any seed that floats can be carried by water.

Hitchhikers

beggar-tick

cocklebur
foxtail

Some seeds have hooks or hairs that catch on people's clothes or animals' fur. These seeds "hitchhike" a ride far from the parent plant.

Seeds Travel

Collect seeds. Look at them carefully with a hand lens. Decide how each one travels. Place seeds into groups. Draw each seed and glue the seed next to your drawing.

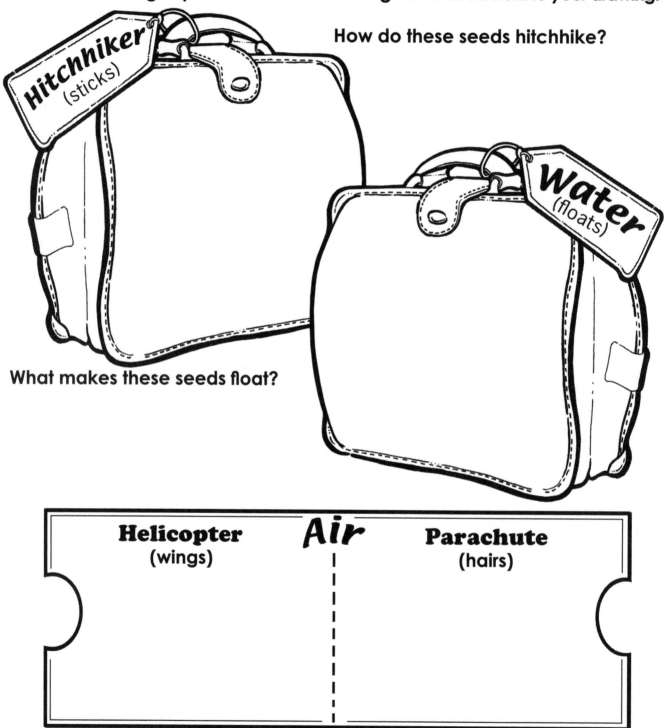

How do these seeds hitchhike?

What makes these seeds float?

How are these seeds carried by the wind?

Find two seeds that can travel by wind.
Draw them.

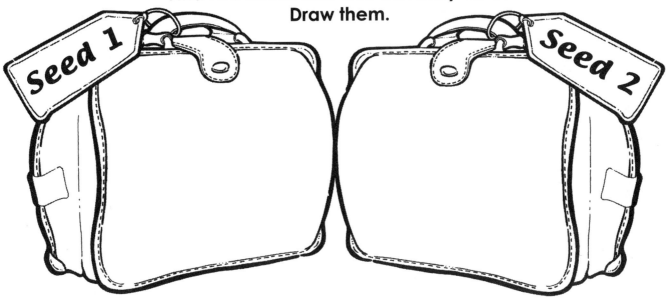

Hold the two seeds over your head and drop them at the same time.
Which one stayed in the air longer? (circle one)

Seed 1   **Seed 2**

How long can each seed stay in the air?

Seed 1	Seed 2
_____ seconds	_____ seconds

Go outside when a breeze is blowing. Drop both seeds.
Which seed travels farther? (circle one)

Seed 1   **Seed 2**

How far did each one travel?

Seed 1	Seed 2
_____ cm	_____ cm

PRIMARILY PLANTS © 2005 AIMS Education Foundation

Connecting Learning

1. What are some ways that seeds travel?
2. Why do they need to travel?
3. How far did your winged seeds travel? Do they need a strong wind?
4. How do the seed cases (the fruit) differ with the various traveling methods?
5. Which do you think is the most efficient mode of travel for the seeds?
6. What is the best kind of weather for airborne seeds to disperse?
7. Can you find some other way in which seeds are dispersed than the ways that have been studied?
8. What are you wondering now?

Observing Bulbs

Topic
Growing bulbs

Key Question
How does a bulb grow?

Learning Goal
Students will plant and observe bulbs as they grow.

Guiding Documents
Project 2061 Benchmark
- People can often learn about things around them by just observing those things carefully, but sometimes they can learn more by doing something to the things and noting what happens.

NRC Standard
- Each plant or animal has different structures that serve different functions in growth, survival, and reproduction. For example, humans have distinct body structures for walking, holding, seeing, and talking.

Math
Measurement
    linear

Science
Life science
    botany
        bulb growth

Integrated Processes
Observing
Comparing and contrasting
Recording data
Interpreting data
Drawing conclusions

Materials
For each group:
    narcissus bulbs (see *Management 1*)
    gravel
    plastic cups, 16 oz
    plastic cups, 10 oz
    ruler

Background Information
Bulbs are a special type of plant. They consist of a round, underground structure that develops in some plants. The bulb is made up of thickened layers of fleshy leaves that hold the stored food. In the center is a large bud scale that produces the new plant. Roots grow from the solid basal plate. The outer scales form a dry and papery covering.

The purpose of the bulb is to store food. When the winter comes, the above ground plant dies, but the bulb with its stored food remains alive underground. When the new growing season begins, the bulb's central bud sends out a shoot, which produces a stem, leaves, and flowers above the ground. Food stored in the bulb starts the new plant's growth.

Onions and garlic are perhaps the food bulbs most familiar to the students. Tulips, daffodils, and narcissuses are garden flower bulbs.

Management
1. Students will plant narcissus bulbs; however, you may want to have various types of bulbs available for students to observe. Suggestions include: onions, garlic, daffodils, tulips.
2. Some flower bulbs are poisonous. Make certain students understand they are not to taste them. Have them wash their hands after handling them.
3. The garden flower bulbs can be bought in the fall in nursery stores. Onions and garlic are available anytime in a grocery store.
4. Fill the 10-oz plastic cups with gravel. Each group will need one cup full.
5. Students should work in groups of three or four.

Procedure
1. Distribute bulbs to students.
2. Tell them to handle the bulbs carefully and observe them with their senses of touch, sight, hearing, and smell.
3. Have students give words that describe the bulbs. Record the words on the board or on a large piece of paper.
4. Discuss with the students that a bulb consists of fleshy food storage leaves. Ask them how this is like the bean seed. [Its cotyledon stores food for the embryo.] Tell them that a stem with leaves and flowers will come from the center of the bulb.

PRIMARILY PLANTS   © 2005 AIMS Education Foundation

5. Tell them that they are going to plant a flower bulb and observe its growth. Distribute a narcissus bulb, cup of gravel, and 16-oz plastic cup to each group. Have them follow the instructions of the student sheet to plant their bulbs. Make sure that students wash their hands after planting the bulbs.

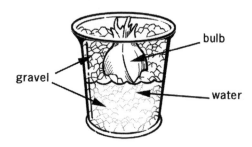

6. Have students keep them watered. Once growth begins, make sure that the plants receive sunlight.
7. Have students notice how the bulb becomes soft as the plant grows. Point out that the plant used the food supply of the bulb to grow the new leaves and roots.

Connecting Learning
1. What is the purpose of the bulb?
2. Where do the roots of the bulb come from?
3. What did the bulbs feel like when you planted them? What did they feel like after they sprouted leaves? Why did the bulb change?
4. What are you wondering now?

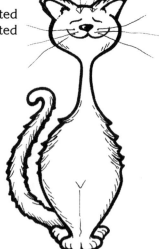

Key Question

How does a bulb grow?

Learning Goal

plant and observe bulbs as they grow.

Observing Bulbs

Use your senses to observe your bulb. Draw and describe what it looks like.

My bulb is a _____

Looks: _____
_____

Smells: _____
_____

Feels: _____
_____

Use your notes to write a description of the bulb.

Connecting Learning

1. What is the purpose of the bulb?

2. Where do the roots of the bulb come from?

3. What did the bulbs feel like when you planted them? What did they feel like after they sprouted leaves? Why did the bulb change?

4. What are you wondering now?

Plants from Cuttings

Topic
Growing plants from cuttings

Key Question
How do cuttings grow?

Learning Goal
Students will observe that plants can be grown by means other than by germination of seeds.

Guiding Documents
Project 2061 Benchmark
- People can often learn about things around them by just observing those things carefully, but sometimes they can learn more by doing something to the things and noting what happens.

NRC Standard
- Each plant or animal has different structures that serve different functions in growth, survival, and reproduction. For example, humans have distinct body structures for walking, holding, seeing, and talking.

Science
Life science
   botany
      plant growth

Integrated Processes
Observing
Comparing and contrasting

Materials
Philodendron or coleus cuttings
Plastic cups, 10 oz
Hand lenses
Potting soil
Carrot top or beet top

Background Information
Plants can reproduce in another way other than by the germination of seeds. They can turn cuttings of themselves into new plants. When a plant reproduces this way, the young plants are genetically identical to the parent.

Management
1. The philodendron plant or coleus plant are perhaps the easiest to grow from cuttings.
2. For the best cutting material, look for healthy, normal tip growth. With a sharp knife cut a four to five inch long stem, making the cut just below a leaf; remove all leaves on the lower half of the stem.
3. Cut one-half inch of the tops of carrots or beets for planting.
4. You may want to mark a calendar with the day the cuttings are started and the day that the roots and leaves are observed growing.
5. This activity can be done individually or in groups.
6. Have students wash their hands after handling the cuttings.

Procedure
Growing philodendron or coleus
1. Take a cutting from a philodendron or a coleus plant. Cut off the stem with a sharp knife just above one of the leaves.
2. Have students follow the directions for preparing the cutting for placing in the cups of water.
3. Provide hand lenses so students can observe the stems before putting the cuttings in water.
4. Invite students to look at their cuttings once a day. In about a week, roots will start to grow from the bottom of the cutting and leaves will sprout from the leaf node.
5. When the roots are about an inch in length, have the students plant the cuttings in soil, pressing down firmly. Tell them to leave enough room so the plant can be watered.

Growing carrots or beets
1. Give each student or group a carrot or beet top. Make sure that a half-inch of the body of the vegetable is left on.
2. Have them plant the tops in a cup in which they cover the carrot or beet with soil so that just the very top part sticks out.
3. In about two weeks, students should see small leaves start poking up. Under the soil, roots will be growing.

Connecting Learning
1. How is growing a plant from cuttings like growing one from seeds? How is it different?
2. From what part of the stem do the roots start to grow?
3. What other fruits or vegetables besides carrots and beets can you find that can grow new tops?
4. What are you wondering now?

Extensions
1. Try growing a begonia or African violet plant from a leaf.
2. Grow a sweet potato vine. Push three toothpicks firmly into the sweet potato at equal distances around the tuber. Put the potato in a cup so that the toothpicks rest on the rim. Keep the water level so it just touches the tip end of the tuber. Sprouts will grow from the potato and in six weeks you will have a vine with attractive foliage.

Plants from Cuttings

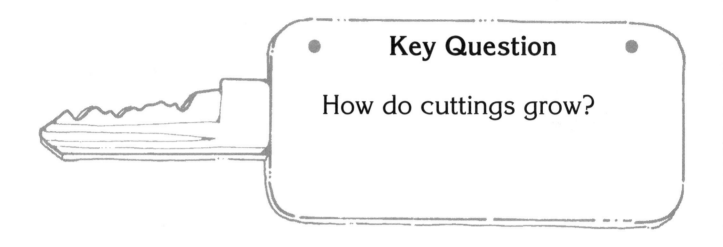

Key Question

How do cuttings grow?

Learning Goal

Students will:

observe that plants can be grown by other means than by germination of seeds.

Plants from Cuttings

You will need:

plastic cups, 10 oz
potting soil
plant cuttings (philodendron or coleus)
vegetable cuttings (carrot or beet top)
water

Do this:

1. Cut a philodendron or coleus stem just below one of the leaf nodes. Trim the lower leaves.
2. Put the cutting in a glass of water.
3. Watch for roots and leaves to grow from the cutting.
4. Put the cutting in a pot and fill with soil. Press down firmly and water when needed.

Another Way

1. Cut one-half inch from the top of a carrot or beet.
2. Plant the top of the vegetable in a pot. Cover with soil so just the very top sticks out.
3. Keep the soil moist and watch for growth.

Plants from Cuttings

Connecting Learning

1. How is growing a plant from cuttings like growing one from seeds? How is it different?

2. From what part of the stem do the roots start to grow?

3. What other fruits or vegetables besides carrots and beets can you find that can grow new tops?

4. What are you wondering now?

Spores: A Special Seed

Topic
Plants with spores

Key Questions
1. Where do spores grow?
2. What do they look like?

Learning Goal
Students will observe spores, a special seed, from which ferns and mosses are reproduced.

Guiding Documents
Project 2061 Benchmarks
- Some animals and plants are alike in the way they look and in the things they do, and others are very different from one another.
- Plants and animals have features that help them live in different environments.

NRC Standard
- Each plant or animal has different structures that serve different functions in growth, survival, and reproduction. For example, humans have distinct body structures for walking, holding, seeing, and talking.

NCTM Standards 2000
- Use tools to measure
- Understand how to measure using nonstandard and standard units

Math
Counting
Measurement
   length

Science
Life science
   botany
      spores

Integrated Processes
Observing
Comparing and contrasting

Materials
Fern fronds with spores
Hand lenses
Rulers

Background Information
Ferns and mosses are two of an important group of green plants that do not form true seeds. They form spores that take the place of the seed in a flowering plant, and develop into a young plant.

The leaves of ferns are called fronds. If you look at the undersides of some fronds, you will see some dark brown spore cases. When the spores are ripe, they blow away on the wind. One fern plant may produce millions of spores.

When a fern spore lands on moist, shady ground and has ideal conditions, it begins to grow. The young plant is shaped like a tiny heart. The little plant is called a prothallus. Soon a small fern begins to grow from the upper surface.

Management
1. This activity is best done during the late summer or fall months when spores form on fern plants. Collect some fern fronds with brown spores on the underside of the leaves.
2. Many floral bunches contain ferns with spore cases.
3. Sword ferns and Boston ferns have spore cases that are easily observed.

Procedure
1. Distribute the fern fronds and student page.
2. Discuss that ferns don't grow from seeds; they grow from spores.
3. Invite students to locate the spore cases. Have them observe them with hand lenses.
4. Tell students to draw the spore cases (brown spots) on the diagram of the fern leaf.
5. Have them select a small leaf and a large leaf to compare their length and number of spore cases. Distribute rulers or other measuring tools.

Connecting Learning
1. What do spores look like? What color are they?
2. Why do you think the spores appear on the underside of the leaves?
3. Why does a plant produce so many spores?
4. How do you think the wind helps the fern?
5. What are you wondering now?

Extensions
1. Make fern prints or fern rubbings.
2. Have the students investigate the number of different ferns around their school and homes. There are nearly 10,000 different kinds of ferns in the world. They could make a collection of the kinds of ferns they find.

PRIMARILY PLANTS

Spores: A Special Seed

Key Questions

1. Where do spores grow?
2. What do they look like?

Learning Goal

Students will:

observe spores, a special seed, from which ferns and mosses are reproduce.

Spores: A Special Seed
Fern Fronds

A fern is a green plant that grows from spores, not seeds.

1. Observe a fern frond. Count its leaves.

    My frond has _____ leaves.

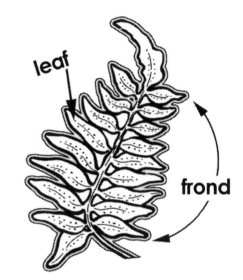

2. Turn the frond over. Look at the spore cases with a hand lens.

3. Pick a leaf about this size. ↴

    Draw the spores on this leaf.

    There are _____ spore cases on the leaf.

4. Pick a small leaf at the top of the frond and a large one at the bottom of the frond.

Compare

Small leaf	Large leaf
length _____	length _____
# spore cases _____	# spore cases _____

PRIMARILY PLANTS © 2005 AIMS Education Foundation

Spores: A Special Seed

Connecting Learning

1. What do spores look like? What color are they?

2. Why do you think the spores appear on the underside of the leaves?

3. Why does a plant produce so many spores?

4. How do you think the wind helps the fern?

5. What are you wondering now?

Plant Needs

Plants are organisms that grow and reproduce their own kind. They need food, air, soil, water, light, and space to grow.

Plants need soil. Water and minerals are taken from the soil through roots. Soil also provides support for the plant and an anchor for the roots to grow in. Decaying plants and animals leave behind minerals in the soil that are essential for future plant growth.

Plants need sunlight in order to grow properly. They use light energy to change the materials—carbon dioxide and water into food substances (sugars). This process of food production is called photosynthesis. Only in light can a green plant make food.

Plants must also have clean air. Green plants take in carbon dioxide from air and use it during photosynthesis to make food. Dirty, smoggy air blocks sunlight that plants must have.

Plants need water. Water is essential to all life on Earth. No known organism can exist without water. Plants use water to carry moisture and nutrients from the roots to the leaves and food from the leaves back down to the roots.

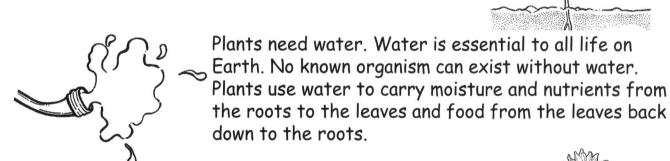

Plants must also have space in order to grow. Plants are found everywhere—deserts, mountains, arctic regions, forests, jungles, oceans, and even in crack of sidewalks of busy cities. If the space is small, the plants will be small and stunted. Big plants need big spaces for their roots and branches.

PRIMARILY PLANTS © 2005 AIMS Education Foundation

Which Soil Works Best?

Topic
Plants: soil

Key Question
Which soils are good for growing plants?

Learning Goal
Students will plant seeds in a variety of soils to see which soils provide the best medium for plant growth.

Guiding Documents
Project 2061 Benchmark
- Plants and animals have features that help them live in different environments.
- Plants and animals both need to take in water, and animals need to take in food. In addition, plants need light.

NRC Standards
- Organisms have basic needs. For example, animals need air, water, and food; plants require air, water, nutrients, and light. Organisms can survive only in environments in which their needs can be met. The world has many different environments, and distinct environments support the life of different types of organisms.
- Ask a question about objects, organisms, and events in the environment.
- Plan and conduct a simple investigation.

Science
Life science
    botany
        plant needs
            soil

Integrated Processes
Observing
Predicting
Comparing and contrasting
Collecting and recording data

Materials
Clear plastic cups, 9-oz
Potting soil
Sand
Playground dirt
Clay
Water
Large bean seeds, 4 per group
Student page

Background Information
Soils provide the water and minerals that plants need. Without soil, plants can be watered, but it becomes difficult to give them the nutrients they gain from the soil. Soil also provides support for plants and their root systems.

Soil usually has three separate layers. The top layer contains minerals and humus, the decayed remains of animals and plants. Humus and minerals are needed by plants for good growth. The second layer contains humus and minerals that have been leached from the topsoil. The third layer consists of rocks that are being broken down to form soil.

A great help to the enrichment of soil is the earthworm. Earthworms burrow through the soil and leave castings of digested leaves and other matter that help enrich the soil. They also improve the soil with their tunneling by making it easier for air and water to soak in.

Management
1. Students should work together in groups of three or four.
2. Each group will need four cups, each one filled with a different type of soil: potting soil, sand, playground dirt, and clay.
3. Soaking the seeds overnight will improve the speed of germination.

Procedure
1. Have students get into groups and distribute the materials. Ask the *Key Question* and discuss students' thoughts.
2. Instruct groups to plant their seeds in their cups. Show them how to put the seeds against the sides of the cups so that they are easily observed.
3. Determine a specific amount of water to be given to all plants and have groups water all four cups.
4. Distribute the student page to each student and have everyone record a prediction about which soil will be best for the seeds.
5. Place the plants in a sunny window where they will all get equal light. Observe the seeds each day

and water them as needed. Be sure that all plants get the same amount of water each time.
6. Once the seeds have sprouted, have students illustrate the results and draw conclusions on their student pages.

Connecting Learning
1. Which soil did you predict would be the best for growing bean plants? Why?
2. How are all the types of soil alike? How are they different?
3. Why do you think one might be better than another for growing bean plants?
4. Which soil grew the best bean plants? Why do you say that? What things are you considering when you say a plant is the "best"? [plant size, color of leaves, number of leaves, speed of germination, etc.]
5. Did the results match your prediction? Why or why not?
6. Does more than one kind of soil produce a good plant? Justify your response.
7. Could soils be combined to make a better growing bed? How?
8. Do you think plants can grow without soil? How?
9. What are you wondering now?

Extensions
1. Create soil mixtures with two or more kinds of soil. Test these and compare the results to the original test.
2. Explore hydroponics—growing plants without soil.

Curriculum Correlation
Language Arts
Write a poem or story about seeds and soil.

Use adjectives to describe the different kinds of soil: rough, sandy, earthy, etc.

Which Soil Works Best?

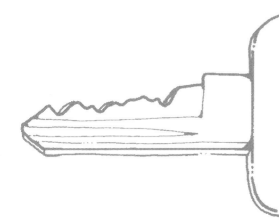

Key Question

Which soils are good for growing plants?

Learning Goal

plant seeds in a variety of soils to see which soils provide the best medium for plant growth.

Which Soil Works Best?

Plant a seed in each cup. Give them the same amount of water.

What do you think will happen?

Watch them grow. Draw what happens to each seed.

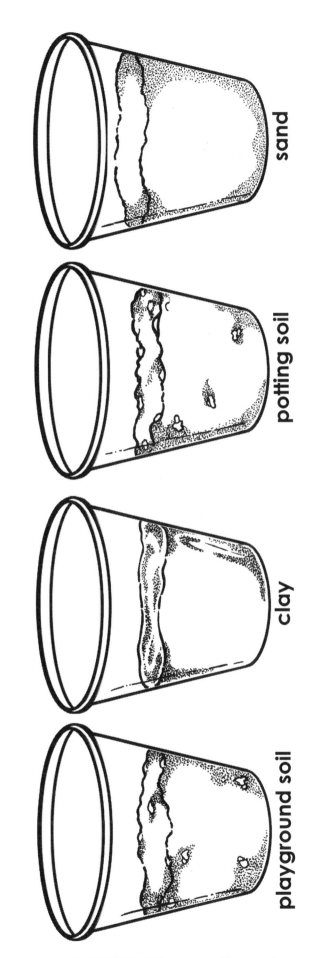

playground soil | clay | potting soil | sand

Which Soil Works Best?

Connecting Learning

1. Which soil did you predict would be the best for growing bean plants? Why?

2. How are all the types of soil alike? How are they different?

3. Why do you think one might be better than another for growing bean plants?

4. Which soil grew the best bean plants? Why do you say that? What things are you considering when you say a plant is the "best"?

5. Did the results match your prediction? Why or why not?

6. Does more than one kind of soil produce a good plant? Justify your response.

7. Could soils be combined to make a better growing bed? How?

8. Do you think plants can grow without soil? How?

9. What are you wondering now?

Plants and Water

Topic
Plants need water

Key Question
Can plants grow without water?

Learning Goal
Students will investigate whether or not a plant needs water to live.

Guiding Documents
Project 2061 Benchmarks
- People can often learn about things around them by just observing those things carefully, but sometimes they can learn more by doing something to the things and noting what happens.
- Plants and animals both need to take in water, and animals need to take in food. In addition, plants need light.

NRC Standard
- Organisms have basic needs. For example, animals need air, water, and food; plants require air, water, nutrients, and light. Organisms can survive only in environments, and distinct environments support the life of different types of organisms.

Science
Botany
  plant needs
    water

Integrated Processes
Observing
Comparing and contrasting
Drawing conclusions
Communicating

Materials
Water
Clear plastic cups
Plants
Potting soil

Background Information
Water is perhaps the most important substance to life on Earth. No known organisms can exist without water.

Plants, like every other living thing, need water in order to live and grow. Water carries the dissolved minerals and nutrients from the soil to the plant, and carries food from the leaves back down to the roots.

The plants get water in several ways—through rainfall, irrigation, and dew. In our homes, we water our plants. Most plants get water from rain. Even in the desert areas, plants would die without moisture.

Management
1. Buy two plants of the same size, variety, and same type pot. You can use plants grown in the previous lesson.
2. This activity may take two weeks to see results. The duration will depend upon the moisture in the soil when the activity begins.

Procedure
1. Ask students if they think plants need water. Ask them how plants get water.
2. Have the students describe what they observe about each plant. Invite them to draw the plants if possible.
3. Put both plants in a sunny spot on the windowsill in view of the class.
4. Assign a child to water one plant. (Do not let children over-water the plant as this is as detrimental to its health as under-watering.) Do not water the other plant.
5. Have the students predict what they think will happen to the plants. (When a plant gets no water, it cannot make food. Without food, the plant will die.)
6. Draw a picture of what happened.

Connecting Learning
1. What did we find out that plants need in this activity?
2. What happens to a plant that doesn't get water?
3. How are the leaves of a plant that gets water different than a plant that doesn't get water?
4. When we need water, what do we do?
5. Can a plant get water on its own?
6. How does a plant get water?
7. What other things need water?
8. What are you wondering now?

Extensions
1. Put radish or marigold seeds (fast growing) on two sponges. Wet one and keep it damp. Keep the other dry and observe the differences that occur.
2. To show that roots take up water from the soil, test with a potato. Peel the bottom half of the potato and cut the end flat so it can stand. Dig a hole in the top part. Stand the potato on its flat end in a dish of water. Watch! The potato will soak up water and fill the hole in the top.

Plants and Water

Key Question

Can plants grow without water?

Learning Goal

Students will:

investigate whether or not a plant needs water to live.

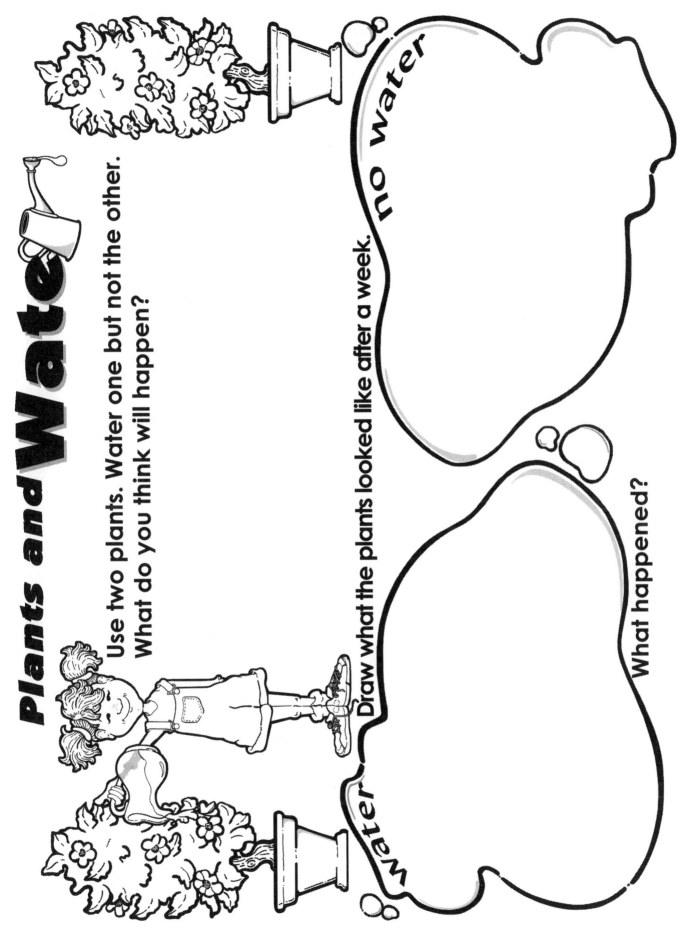

Plants and Water

Connecting Learning

1. What did we find out that plants need in this activity?

2. What happens to a plant that doesn't get water?

3. How are the leaves of a plant that gets water different than a plant that doesn't get water?

4. When we need water, what do we do?

5. Can a plant get water on its own?

6. How does a plant get water?

7. What other things need water?

8. What are you wondering now?

Blue Ribbon Crops

Topic
Plants and water

Key Question
How can water affect the growth of lima bean plants?

Learning Goals
Students will:
• identify the effect that water can have on the germination of plant seeds, and
• identify ways that water can affect plant growth.

Guiding Documents
Project 2061 Benchmarks
• Most living things need water, food, and air.
• Plants and animals both need to take in water, and animals need to take in food. In addition, plants need light.

NRC Standard
• Organisms have basic needs. For example, animals need air, water, and food; plants require air, water, nutrients, and light. Organisms can survive only in environments in which their needs can be met. The world has many different environments, and distinct environments support the life of different types of organisms.

NCTM Standards 2000*
• Pose questions and gather data about themselves and their surroundings
• Represent data using concrete objects, pictures, and graphs
• Count with understanding and recognize "how many" in sets of objects

Math
Graphing
Number sense
Counting

Science
Life science
    plants

Integrated Processes
Observing
Communicating
Collecting and recording data
Inferring
Predicting
Controlling variables

Materials
For each group of students:
  1 set of *Plot Markers*
  40 lima bean seeds
  plant box (see *Management 2*)
  trash bag, kitchen-size
  planting soil (see *Management 2*)
  small clay flowerpot, 2" in diameter
  string
  tape
  4 craft sticks
  water
  1 set award ribbons, included

For the class:
  chart paper

Background Information:
Plants are composed of about 70% water. Water is essential to them in many ways. It is necessary in the beginning stages of the plant's life to soften the seed coat. This allows the root and stem within to push open the seed coat. Water also dissolves minerals in soil. Plant roots pick up the water and dissolved minerals and transport these materials up the stem to the leaves. All stages of a plant's life require water. Through the experiences involved in this activity, students should develop an understanding that plants need water, but too much water can be harmful.

Students will also gain experience with controlling and manipulating variables. In this activity the controlled variables are the number of seeds planted per section and the type of soil used for all sections. The manipulated variable is the amount of water the various sections will get due to their distance away from the water source.

Management
1. *Part One* of this activity will take two weeks to complete. *Part Two* is extended observations.
2. Lids from boxes of copy paper can be used to make the planting boxes. You will need to line the boxes with plastic. Large kitchen-size trash bags work well. Place box lid in the bag and close it with a twist tie. Place soil in the box—soil from the yard is best; however, potting soil will do. You will need one box per group.
3. Each group will need 20 beans. Lima bean seeds are used because of their rapid germination rate. They are also easier to count for the math connection of this

PRIMARILY PLANTS   97   © 2005 AIMS Education Foundation

activity. You can substitute other seeds. Radish and tomato seeds germinate rapidly, but they are small.

4. The ribbon page can be copied on different colored paper: blue—first place; red—second place; white—third place; and yellow—fourth place. Each group will need four ribbons, one for each place. There is a space for writing the place on each ribbon.
5. Recommended group size for this activity is four students per group.
6. Make certain that the flowerpots have a drain hole in the bottom.
7. The students can make their own Country Fair Journals or use notebook paper to record their daily observations.

Procedure
Part One
1. Ask the *Key Question*.
2. Have the students put the flowerpot in the corner of the plant box and add soil to approximately 2 inches deep. Make sure the soil is firmly packed in the box. Show them how to use the string to make four diagonal rows about 3 inches apart. Instruct them to attach the string to the sides of the box with tape to keep it secure.

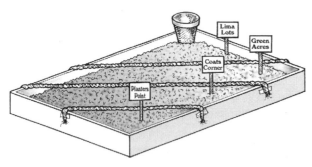

3. Tell the students that they are going to plant some seeds for the Country Fair. Explain that each group will award a ribbon to each of their four plots according to the number of beans that sprout.
4. Give each group of students 20 lima beans and let the students divide the seeds into four equal groups.
5. Have each group of students lay the seeds on top of the soil so that they are as evenly spaced as possible. Ideally, the seeds should have 3-4 inches of space between them. Each section needs five seeds total.
6. Once groups have a good spacing plan, instruct them to push the seeds into the soil with their fingers and to cover the holes.
7. Have the students make markers by gluing the *Plot Markers* to the ends of craft sticks. Tell them to label the plot closest to the watering pot *Lima Lots*, the second plot *Green Acres*, the third plot *Coats Corner*, and the fourth plot *Planters Point*.
8. Instruct the students to fill the small clay pot with water. Each time it empties, have them refill it.
9. Have students predict which area will have the fastest germination rate. Write their predictions on a piece of chart paper, which can be accessed later for use. [The seeds nearest the water will sprout first. Another row may germinate. The seeds farthest away from the water source usually do not sprout due to lack of water.]
10. Have students keep a daily record of the results in their journals, including a reading on the degree of dampness in each section (very wet, wet, damp, dry).
11. Have students use the picture graph to record the number of seeds that germinate in each section.
12. At the end of two weeks of observation, have each group award the ribbons to the four sections based on germination rate.
13. Culminate this part by having the students make a bar graph to compare their group's results to the results of the other groups.

Part Two
1. Have the students continue watering and observing the plant boxes for two additional weeks.
2. Record the class observations from the second two weeks on chart paper.

Connecting Learning
1. What did you observe?
2. Why did you only put water in the pot and not over all of the plant box?
3. What row had the most seeds sprout?
4. Which seeds sprouted first? [Those nearest the water pot.] Why do you think they sprouted first?
5. Did the first row end up having plants that grew best? [No.] Why not? [They got too much water.]
6. What did you learn about the effect that water has on a plant? [Receiving too much water can be as harmful to a plant as not receiving enough water.]
7. How can we use this information to help us care for our house/classroom plants?
8. What are you wondering now?

Evidence of Learning
1. Listen as students offer reasons for the seed-sprouting rate.
2. Listen for explanations on how water affects plant growth.
3. Read the daily journals for observation.

Extensions
1. Give each group of students a second set of the plot markers and ask them to put them in order from best growing area to worst growing area. Compare the results.
2. Have the students determine what fraction of the seeds germinated. They can also determine the decimal value and then convert this number into a percentage.

* Reprinted with permission from *Principles and Standards for School Mathematics,* 2000 by the National Council of Teachers of Mathematics. All rights reserved.

Blue Ribbon Crops

Key Question

How can water affect the growth of lima bean plants?

Learning Goals

Students will:

- identify the effect that water can have on the germination of plant seeds, and
- identify ways that water can affect plant growth.

Plot Markers

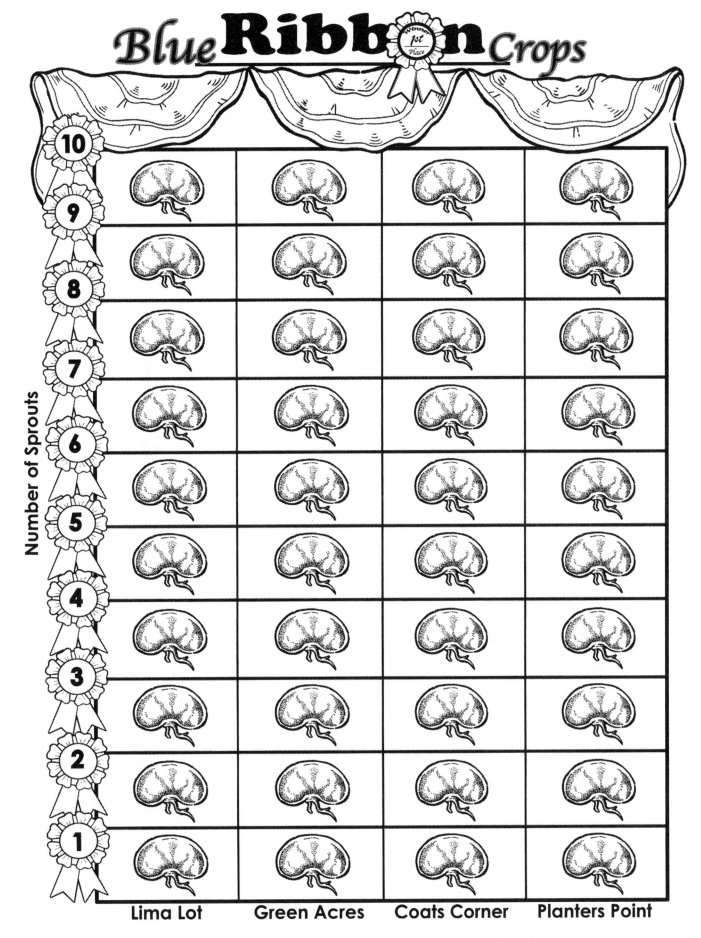

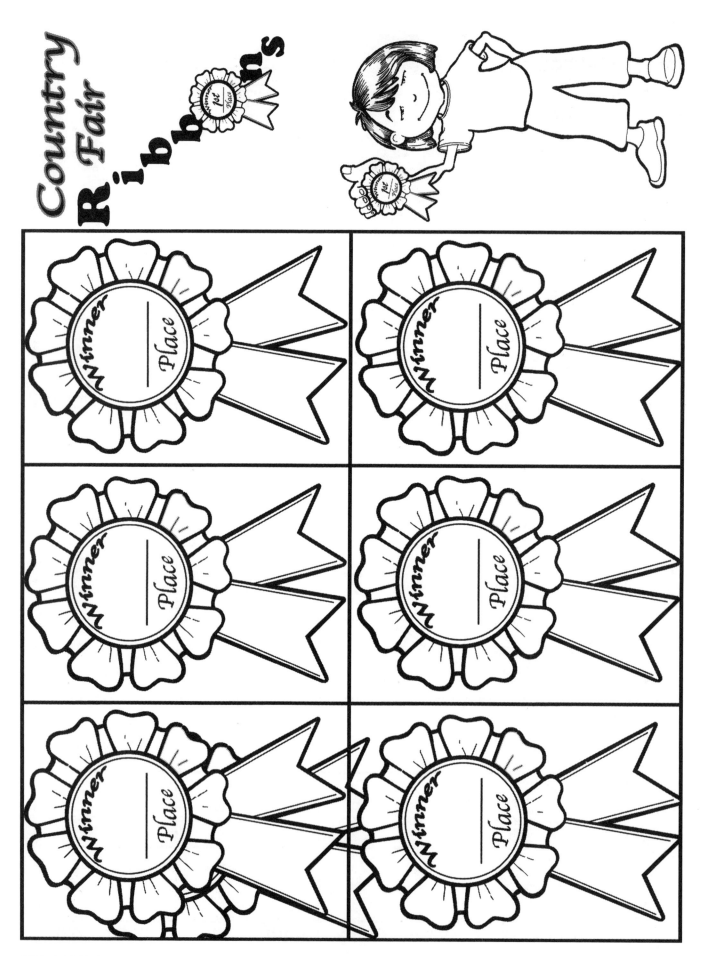

Blue Ribbon Crops

Which section had the best sprouting rate?

Connecting Learning

1. What did you observe?

2. Why did you only put water in the pot and not over all of the plant box?

3. What row had the most seeds sprout?

4. Which seeds sprouted first? Why do you think they sprouted first?

5. Did the first row end up having plants that grew best? Why not?

6. What did you learn about the effect that water has on a plant?

7. How can we use this information to help us care for our house/classroom plants?

8. What are you wondering now?

Topic
Plants need sunlight

Key Question
Do plants need sunlight in order to grow?

Learning Goal
Students will be able to prove that a plant needs light in order for it to develop correctly.

Guiding Documents
Project 2061 Benchmarks
- People can often learn about things around them by just observing those things carefully, but sometimes they can learn more by doing something to the things and noting what happens.
- Plants and animals both need to take in water, and animals need to take in food. In addition, plants need light.

NRC Standard
- Organisms have basic needs. For example, animals need air, water, and food; plants require air, water, nutrients, and light. Organisms can survive only in environments, and distinct environments support the life of different types of organisms.

Science
Botany
   plant needs
      sunlight

Integrated Processes
Observing
Comparing and contrasting
Drawing conclusions
Communicating

Materials
Sunflower seeds (any alternative seed will do)
Potting soil
Plastic cups, 9- or 10 oz.
Masking tape
Boxes
Aluminum foil
House plant
Two plants the same size

Background Information
Plants need food and can make their own food. But they need things from which to make their food. Green plants need carbon dioxide from the air, water and minerals from the soil, and energy from the sun.

Only in light can a green plant make food. The process of food-making is called photosynthesis. In photosynthesis, the carbon dioxide and water are changed to carbohydrates and oxygen. Food can only be made in the presence of chlorophyll. Chlorophyll is the substance responsible for a plant's green color.

When a green plant is deprived of sunlight, it soon loses it chlorophyll. It cannot make food, so it dies.

Management
1. Use sunflower seeds or any fast growing seeds. Soak the sunflower seeds overnight.
2. Remember, the first leaves are not the true leaves; wait for the second set of leaves.
3. Get two plants of the same size.
4. Be sure that the plant is healthy and receives good care during the time one leaf is covered by aluminum foil.

Procedure
1. Tell students that they are going to investigate whether plants need sunlight in order to grow and be healthy. Discuss how they could investigate this by letting one plant have sunlight and keeping the other plant under a box so it will be in the dark.
2. Distribute two cups to each group of students. Have them write their names on masking tape to label their cups.
3. Have students fill the cups half full of potting soil.
4. Give them three sunflower seeds to plant in each cup, and tell them to put the cups near the window. Have the students water the cups well, but caution them not to over-water them.
5. Direct students to cover one of their cups with the cardboard box. Tell them not to lift the box until they are told to do so.
6. When the first true leaves appear on the plants that are uncovered, have students remove the box from the other cup.

7. Discuss with the students what happened to the plants. Why did the sunflower seeds under the box not produce healthy plants? Do plants need light to grow? Do they need light to remain healthy? When a plant gets no sunlight, it cannot make food.
8. Tell students that they will check to see if light makes a difference to plants that are already growing. Show them the two plants that are of the same size. Cover one with a box or place it in a closet to block out all light. After one week without light, bring the plant out and compare it to the one that received sunlight. (The plant that did not receive sunlight will lose some of its green color. It will not look as healthy as the other plant. Make sure it now receives sun and watch for it to return to health.)
9. Ask students what they think will happen if they just cover one leaf with aluminum foil so it doesn't get light. Have students observe the leaves of a healthy houseplant. Cover one of its leaves with aluminum foil. Leave it covered for a week, then uncover and observe the changes. (The leaf should be pale compared to other leaves on the plant.)

Connecting Learning
1. What happens to a plant that has no sunlight?
2. Do plants need light in order to grow? How do you know?
3. What color are the leaves of the plant that receives sunlight?
4. What color are the leaves of the plant that doesn't receive sunlight?
5. What are you wondering now?

Extensions
1. Put a potato in a dark, warm spot for several weeks. What happens to the potato? If left in the dark for several months, will the potato die? Try it.
2. Lay three or four seeds in the bottom of a shallow bowl and place a wet sponge over them. Keep them damp. When the seeds sprout, watch to see what happens to them. Turn the dish. Do the seedlings turn toward the light?
3. Put a rock or a board on a patch of grass, leave it for two weeks. Take the rock off and observe. What has happened to the grass? Is it a different color from the surrounding grass? Has the grass died? Now leave the rock or board off the grass. Does the green color return to the grass?
4. Write a plant log about one of the experiments.
5. Let the students make up an activity of their own with plants and sunshine.

Key Question

Do plants need sunlight in order to grow?

Learning Goal

Students will:

be able to prove that a plant needs light in order for it to develop correctly.

Connecting Learning

1. What happens to a plant that has no sunlight?

2. Do plants need light in order to grow? How do you know?

3. What color are the leaves of the plant that receives sunlight?

4. What color are the leaves of the plant that doesn't receive sunlight?

5. What are you wondering now?

Plants & Space

Topic
Plants need space

Key Question
Do plants need space in order to develop correctly?

Learning Goal
Students will understand that plants grow in many places and need space.

Guiding Documents
Project 2061 Benchmarks
- *People can often learn about things around them by just observing those things carefully, but sometimes they can learn more by doing something to the things and noting what happens.*
- *Plants and animals both need to take in water, and animals need to take in food. In addition, plants need light.*

NRC Standard
- *Organisms have basic needs. For example, animals need air, water, and food; plants require air, water, nutrients, and light. Organisms can survive only in environments, and distinct environments support the life of different types of organisms.*

Science
Botany
  plant needs
    space

Integrated Processes
Observing
Comparing and contrasting
Drawing conclusions
Communicating

Materials
Two large pots
Radish seeds or any fast growing seeds
Water
Potting soil

Background Information
Plants are found everywhere—gardens, lawns, forests, flowerpots, hillsides, or in the water. Plants can even be found in cracks in the sidewalk and cracks in tar. If the spaces are small, plants will be small. If the spaces are large, the plants tend to be large. Plants can grow up, down, around a tree, across the soil, in the water, and in almost all kinds of places.

Management
1. Use small seeds that grow fast.
2. Prepare two pots with soil.

Procedure
1. Have the students look out the window and answer the question, "Where do plants grow?" (List answers on the board.)
2. Ask "Do all plants need the same amount of space to grow in?" [no] Explain.
3. Compare the amount of space a tree needs for growth against that of a grass plant.
4. Prepare two pots with soil. Plant both with radish seeds. Give both pots the same amount of water and sunlight. Leave the pots alone for 12 days.
5. In one pot, leave all the plants that are growing.
6. In the second pot, thin the baby plants until only three or four are left.
7. Give the pots water and sunlight as before for another 10 days. Then compare the growth and the roots of the plants in both pots.

Connecting Learning
1. What was done differently to the plants in the two pots? [One pot had a lot of plants, the other pot only had a few.] What things did we keep the same? [same size pot, same amount of water, same amount of sunshine]
2. Why are the plants in the pot that has been thinned so much larger and healthier looking? [They have fewer plants to share the water and nutrients in soil.]
3. What happens to trees when there are too many trees crowded together? [They are spindly.] Can you think of a place where the trees are not big around because there are so many of them? If so, where?
4. How do you feel when you are crowded on a bus or in a room and can't move. Do you like it? Do you think it is always healthy?
5. What other things don't like to be crowded?

Plants & Space

Key Question

Do plants need space in order to develop correctly?

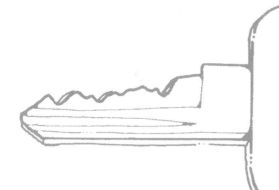

Learning Goal

Students will:

understand that plants grow in many places and need space.

Plants & Space

Plant 20 radish seeds in each cup. Cover with soil.

Give the same amount of water and light to both cups.

After 12 days, thin plants from one cup so there are only 4 left. Leave the other cup alone.

—— What do you think will happen? ——

After 3 weeks, draw what happened to the plants.

Why do you think this happened?

Connecting Learning

1. What was done differently to the plants in the two pots? What things did we keep the same?

2. Why are the plants in the pot that has been thinned so much larger and healthier looking?

3. What happens to trees when there are too many trees crowded together? Can you think of a place where the trees are not big around because there are so many of them? If so, where?

4. How do you feel when you are crowded on a bus or in a room and can't move. Do you like it? Do you think it is always healthy?

5. What other things don't like to be crowded?

Patchwork Planting

Topic
Plants need space to grow

Key Question
How does plant spacing in soil affect growth?

Learning Goals
Students will:
- identify the effects that spacing in soil can have on seed germination, and
- identify ways that space can affect plant growth.

Guiding Documents
Project 2061 Benchmarks
- People can often learn about things around them by just observing those things carefully, but sometimes they can learn more by doing something to the things and noting what happens.
- Describing things as accurately as possible is important in science because it enables people to compare their observations with those of others.

NRC Standard
- Organisms have basic needs. For example, animals need air, water, and food; plants require air, water, nutrients, and light. Organisms can survive only in environments, and distinct environments support the life of different types of organisms.

NCTM Standards 2000
- Pose questions and gather data about themselves and their surroundings
- Recognize and apply mathematics in contexts outside of mathematics
- Count with understanding and recognize "how many" in sets of objects

Math
Graphing
Counting
One-to-one correspondence
Measuring

Science
Life science
    plants

Integrated Processes
Observing
Collecting and recording data
Analyzing data
Comparing and contrasting
Communicating
Controlling variables
Inferring
Predicting

Materials
For each group of students:
    Patchwork Planting Journal
    78 lima bean seeds
    plant box (see *Management 2*)
    potting soil
    string
    water
    masking tape
    trash bag, kitchen-size
    craft stick
    3 x 5 card
    ruler

Background Information
Plants need space in order to survive in an environment. The amount of space needed by a plant to perform well is a function of its size. Both the visible parts of the plant as well as its root system need space. The smaller a plant is when it matures, the smaller the space it needs. Plants use the space available and utilize the sunlight, water, nutrients, carbon dioxide, and oxygen the space contains. If the spacing between plants is too small, a plant uses most of its energy in competing for space and other resources rather than producing a good healthy plant.

This activity will engage the students in an experience that requires the use of variables. The students will investigate spacing (manipulated variable). They will examine how spacing affects the rate of growth of plants (responding variable). The students will control the aspects of the investigation that they can by using the same type of seeds, using the same type of soil, and by keeping the sunlight and water as equal as they are able to.

Management

1. *Part One* of this activity will take two weeks to complete. *Part Two* is extended observations.
2. Lids from boxes of copy paper can be used to make the planting boxes. You will need to line the boxes with plastic to prevent water from leaking through. Large kitchen-size bags work well. Place the box lid in the bag and close it with a twist tie. Place soil in the box—soil from the yard is best; however, potting soil will do. You will need one box per group.
3. Each group will need 78 beans. Lima bean seeds are recommended because of their size and relatively rapid germination rate. They are also easier to count for the math connection of this activity. You can substitute other seeds. Radish and tomato seeds germinate rapidly, but are very small and could be difficult for children to count and sort.
4. Use the *Patchwork Planting Journal* to keep a written record of students' observations. Use plain copy paper as needed to add more pages to the journal.
5. Recommended group size for this activity is four students per group.

Procedure

Part One

1. Explain how to mark off four equally spaced horizontal rows with string along the long side of the box lid. Instruct students to hold the string in place with the masking tape.
2. Direct the students to create a grid system by dividing the planting box into three equally spaced sections along the short side of the box.

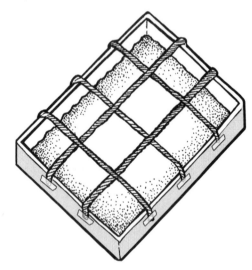

3. Have students make a marker for the first section of the grid by writing their group's name on a 3 x 5 card and taping it to a craft stick. For example, *Tiger Patch*.
4. Have students plant one seed in the first patch, two in the second and so on, until they have planted 12 seeds in the last patch. Instruct them to space the seeds as equally as possible within each patch of soil.

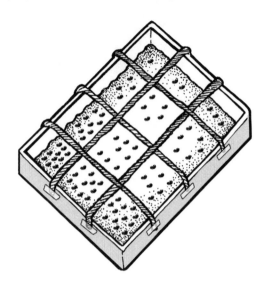

5. Put the plant boxes in a spot where they will get plenty of sun. Water the patch daily, or as needed.
6. Ask students to make daily observations and record what they observe in their *Patchwork Planting Journal*. Encourage them to measure the height of the plants after germination.

Part Two

1. Direct students to observe the plant boxes for two additional weeks.
2. Instruct students to record the height of the plants in each patch over the next two weeks. Have students continue the daily written record of the plant growth.

Connecting Learning

1. What did you observe?
2. Did all seeds sprout? Explain.
3. How did the plants look that had more growing room? How does that compare to those that were crowded?
4. How does spacing in soil help a plant?
5. In which patches were the healthiest plants found at the end of the investigation?
6. Why do you think space is necessary for growing healthy plants?
7. How can we use this information to help us care for our house/classroom plants?
8. Why was it important that we watered the patches all the same?
9. What are you wondering now?

* Reprinted with permission from *Principles and Standards for School Mathematics,* 2000 by the National Council of Teachers of Mathematics. All rights reserved.

Key Question

How does plant spacing in soil affect growth?

Learning Goals

Students will:

- identify the effects that spacing in soil can have on seed germination, and
- identify ways that space can affect plant growth.

Connecting Learning

1. What did you observe?

2. Did all seeds sprout? Explain.

3. How did the plants look that had more growing room? How does that compare to those that were crowded?

4. How does spacing in soil help a plant?

5. In which patches were the healthiest plants found at the end of the investigation?

6. Why do you think space is necessary for growing healthy plants?

7. How can we use this information to help us care for our house/classroom plants?

8. Why was it important that we watered the patches all the same?

9. What are you wondering now?

What Do Plants Need to Grow?

Topic
Plant needs

Key Question
What do plants need to grow?

Learning Goal
Students will understand that in order to grow healthy plants, soil, water, light, and air must be provided.

Guiding Documents
Project 2061 Benchmarks
- People can often learn about things around them by just observing those things carefully, but sometimes they can learn more by doing something to the things and noting what happens.
- Plants and animals both need to take in water, and animals need to take in food. In addition, plants need light.

NRC Standard
- Organisms have basic needs. For example, animals need air, water, and food; plants require air, water, nutrients, and light. Organisms can survive only in environments in which their needs can be met. The world has many different environments, and distinct environments support the life of different types of organisms.

Science
Botany
  plant needs

Integrated Processes
Observing
Comparing and contrasting
Drawing conclusions
Communicating

Materials
For each group:
  4 milk cartons, half-pint (see *Management 1*)
  bean, radish, or corn seeds
  potting soil mixture
  plastic bag, gallon size with twist tie
  cardboard box
  water

Background Information
Plants require sunlight, water, soil, and air in order to grow and be healthy. Energy received from the sun is used to convert carbon dioxide and water into food. When plants do not receive the things they need to live and grow, they will either die or be stunted in their growth.

Management
1. Save the half-pint milk cartons from the students' lunches, rinse them and cut the tops off. If milk cartons are not available, use 9-oz. plastic cups.
2. Use fast growing seeds such as radish, corn, or bean seeds.
3. A shoebox works well for the cardboard box.
4. Students should work in groups of four.

Procedure
1. Ask the *Key Question* and state the *Learning Goal*.
2. Tell students that they are going to find out what happens when a plant does not get its needs met. Inform them that this will take some time because they first have to grow some plants so they can do the experiment.
3. Distribute one milk carton to each student.
4. Have them fill the carton with a soil mixture.
5. Supervise as they plant the seeds in the milk cartons, four or five seeds per carton is enough. Watch as they dampen the soil. Caution them not to over-water the seeds.
6. After the seedlings sprout, remove all but the healthiest plant so there is only one per container. Assign each student in the group a letter: A, B, C, or D. This letter will tell them under what conditions their plants will grow.
   - Condition A—Plant has soil, water and light, but does not have air. Put these plants in a plastic bag and use a twist tie to close it.
   - Condition B—Plant has soil, water, and air, but no light. Put this plant under a cardboard box.
   - Condition C—Plant has soil, air, and light, but no water. Do not water these plants.
   - Condition D—This is the control group. The plants have soil, air, light, and water.

7. Invite students to measure each week how much the plants have grown in each environmental condition.
8. Have students record the measurements on the graphs.
9. After several weeks, compare the graphs. Are there differences in rate of growth of the different plants in the separate condition?

Connecting Learning
1. What does each plant need in order to grow? [soil, air, light, and water]
2. What did the plants look like in each of the conditions and what need was lacking in each one?
3. Why was it necessary to have some plants that had all their needs met? [We needed to have them so we could compare the others to them.]
4. Which plants grew the least? What need were they missing?
5. If you have plants at home, how are their needs met?
6. How are the needs of outdoors plants met?
7. What are you wondering now?

Key Question

What do plants need to grow?

Learning Goal

Students will:

understand that in order to grow healthy plants, soil, water, light, and air must be provided.

 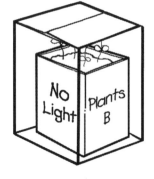  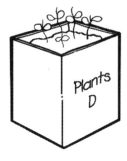

What Do Plants Need to Grow?

You will need:
4 milk cartons
soil
radish, bean, or corn seeds
scissors

Do this:
1. Cut off the top of the milk cartons to make planters.

2. Decorate with roving or paper.

3. Fill the cartons with soil.

4. Plant the seeds in the soil. Dampen the soil.

5. Wait. After the seeds sprout, divide the cartons into 4 groups to test growing conditions.

Has: soil, water, light
No Air

Has: soil, water, air
No Light

Has: soil, air, light
No Water

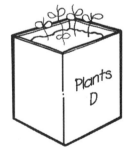
Has soil, air, light, and water

6. Watch to see which plants grow best. What do plants need to grow?

PRIMARILY PLANTS © 2005 AIMS Education Foundation

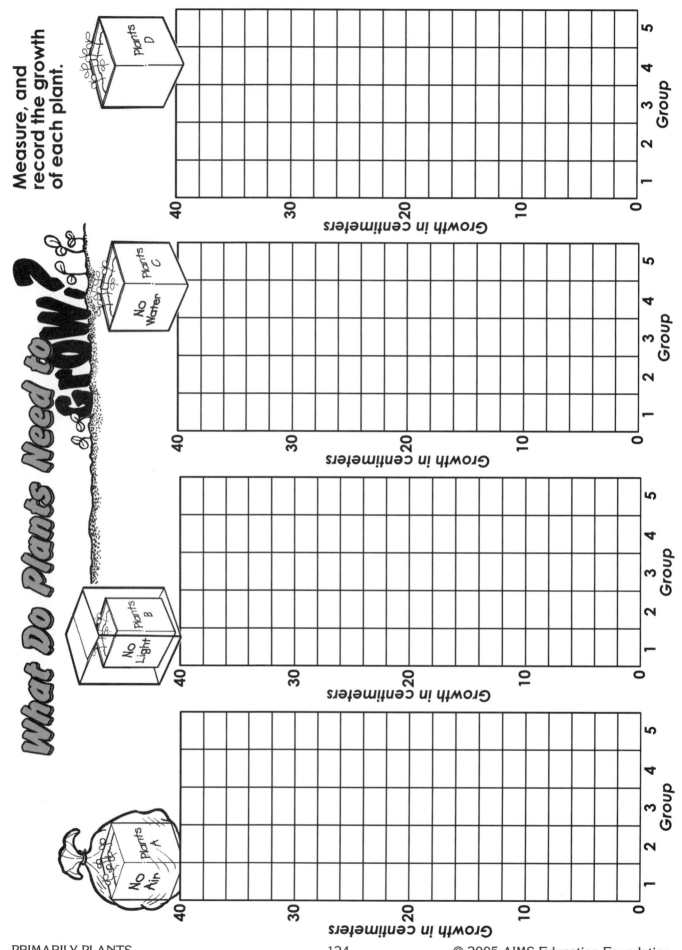

Connecting Learning

1. What does each plant need in order to grow?

2. What did the plants look like in each of the conditions and what need was lacking in each one?

3. Why was it necessary to have some plants that had all their needs met?

4. Which plants grew the least? What need were they missing?

5. If you have plants at home, how are their needs met?

6. How are the needs of outdoors plants met?

7. What are you wondering now?

What Temperature is Best?

Topic
Plant needs: temperature

Key Question
How do plants react to temperature extremes?

Learning Goal
Students will put plants in three different environments to see how they respond to extremes of temperature.

Guiding Documents
Project 2061 Benchmark
- *Plants and animals have features that help them live in different environments.*

NRC Standards
- *Organisms have basic needs. For example, animals need air, water, and food; plants require air, water, nutrients, and light. Organisms can survive only in environments in which their needs can be met. The world has many different environments, and distinct environments support the life of different types of organisms.*
- *Ask a question about objects, organisms, and events in the environment.*
- *Plan and conduct a simple investigation.*

Science
Life science
    botany
        plant needs
            temperature

Integrated Processes
Observing
Predicting
Comparing and contrasting
Collecting and recording data

Materials
Three similar houseplants (see *Management 2*)
Heater (see *Management 3*)
Freezer or refrigerator (see *Management 4*)
Crayons or colored pencils
Student page

Background Information
Different kinds of plants live in different places. The temperature, length of season, and amount of rainfall help to determine the kinds of plants that will grow in a given location. Proper conditions for plant growth can be created on farms and in greenhouses. This activity will allow students to see how one kind of plant reacts to varying temperatures.

Management
1. This activity takes five days, and should be started on a Monday.
2. You will need three houseplants for this activity. Select three healthy plants of the same variety that are as similar in size as possible. Be sure they are all equally well watered before beginning the activity.
3. If possible, this activity should be done during hot weather. If it is done in the winter, you will need a heater by which to place one of the plants.
4. To simulate a cold environment, use a freezer or refrigerator. If this is not possible, use an ice chest full of ice.

Procedure
1. Show students the plants and discuss how they are essentially the same.
2. Distribute the student page and crayons or colored pencils. Have each student draw a picture of each plant.
3. Tell students that you will be placing one plant in a hot environment, one in a cold environment, and one in a moderate environment. Ask them to predict which plants will be healthy at the end of five days.
4. Have them record their predictions.
5. Put the plants in their respective environments. Observe them every day for five days.
6. At the end of five days, have students once again draw a picture of each plant and record their observations.
7. Discuss the results.

Connecting Learning
1. What happened to the plant that was in the hot environment? Is this what you predicted? Why or why not?
2. What happened to the plant that was in the cold environment? Is this what you predicted? Why or why not?
3. What happened to the plant that was in the moderate environment? Is this what you predicted? Why or why not?
4. Do you think the results would have been the same if we had used a different kind of plant? Why or why not?
5. Are there plants that do well in hot environments? Give some examples. [cacti, succulents, etc.]
6. Are there plants that do well in cold environments? Give some examples. [snow plants, pine trees, etc.]
7. What are you wondering now?

Extension
Repeat the activity using different kinds of plants—some that are more suited to cold and some that are more suited to heat. Compare the results.

What Temperature is Best?

Key Question

How do plants react to temperature extremes?

Learning Goal

Students will:

put plants in three different environments to see how they respond to extremes of temperature.

What Temperature is Best?

Get three plants that are alike.
Give each plant the same amount of water.
Draw each plant.

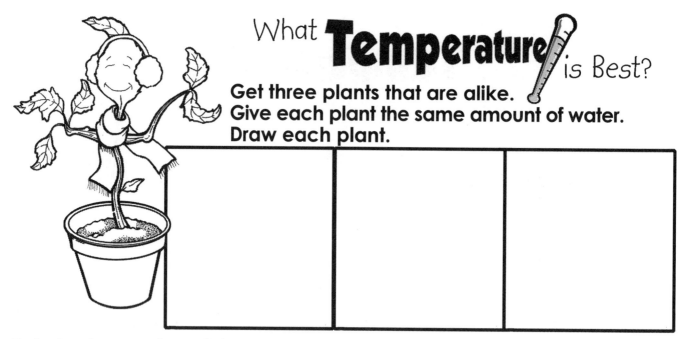

Put plant one where it is hot—in the sun.
Put plant two where it is cold—in the freezer.
Put plant three where it is moderate—in the classroom.

Which plants do you think will be healthy after five days?

Observe the plants every day. Draw the plants after five days.

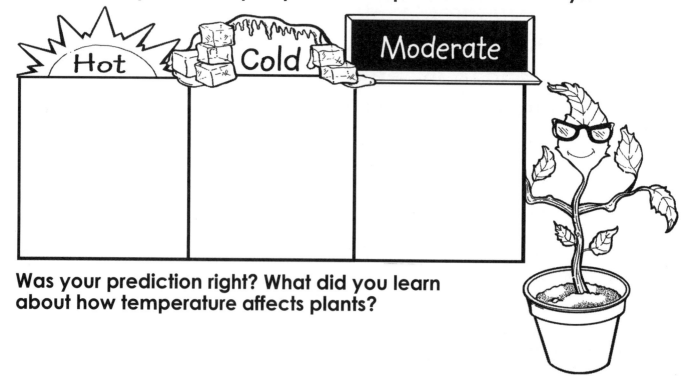

Was your prediction right? What did you learn about how temperature affects plants?

Connecting Learning

1. What happened to the plant that was in the hot environment? Is this what you predicted? Why or why not?

2. What happened to the plant that was in the cold environment? Is this what you predicted? Why or why not?

3. What happened to the plant that was in the moderate environment? Is this what you predicted? Why or why not?

4. Do you think the results would have been the same if we had used a different kind of plant? Why or why not?

5. Are there plants that do well in hot environments? Give some examples.

6. Are there plants that do well in cold environments? Give some examples.

7. What are you wondering now?

What Do Plants Need?

Use the booklet "What Do Plants Need?" as a culminating activity for the section on plant needs.

Have students color or decorate the booklet as they wish. Place on the class library shelf for free reading or send home for students to read to their families.

It is also possible to enlarge the booklet to 11" x 17" and use as a class big book. Read each page as the class participates in an activity that matches.

Plants need soil.

Plants need soil and water.

Plants need soil, water, and light.

Plants need soil, water, light, and space to grow.

People need plants for food.

People need plants for clothing.

People need plants for shelter.

Reaching Up Toward the Sun

Topic
Growth of a sunflower plant

Key Question
How does a sunflower seed grow into a mature plant?

Learning Goals
Students will:
- germinate sunflower seeds, and
- plant the seeds and observe the growth cycle.

Guiding Documents
Project 2061 Benchmarks
- *Change is something that happens to many things.*
- *A lot can be learned about plants and animals by observing them closely but care must be taken to know the needs of living things and how to provide for them in the classroom.*

NRC Standard
- *Plants and animals have life cycles that include being born, developing into adults, reproducing, and eventually dying. The details of this life cycle are different for different organisms.*

Science
Life science
    plants

Integrated Processes
Observing
Predicting
Collecting and recording data
Comparing and contrasting
Communicating

Materials
For each group:
    sunflower seeds
    transparent plastic cups, 8 oz.
    potting soil to fill the cups
    zipper-type plastic bags, pint size
    water
    paper towel

For the class:
    spray bottle with water
    plastic wrap
    chart paper, optional

Background Information
(See Information sheet.)
  To germinate means to begin to grow. A sunflower seed, like all seeds, needs to absorb water until it swells and bursts its seed coat. It will take seven to 12 days for a sunflower seed to germinate. A seed contains food to support the tiny embryo. The root tip usually emerges out of the seed first. It grows rapidly, absorbing water and minerals from the soil and anchors the developing seedling. Then the young stem and leaves emerge from the seed and push their way through the surface of the soil into sunlight. The leaves turn green when light reaches them and the plant begins to manufacture its own food. Sunflower plants will mature and start to flower in 80 to 90 days.

Management
1. Get some sunflower seeds from a nursery or from a seed catalog.
2. For the first part of the lesson, each group of students will need four sunflower seeds, a plastic bag, and a paper towel. You will also need a spray bottle filled with water that the students can share.
3. For the second part of the lesson, each group of students will need a clear plastic cup, moistened soil to fill the cup, and a permanent marker to label the cups. Seeds from the first part of the lesson will also be used. Have some extra seeds on hand in case of accidents when students transfer the seeds from the seed bags to the cups of dirt.
4. The students will keep a journal throughout this activity. Copy the pages and have students cut them apart, order them, and staple along the lefthand side. You will need to give students several copies of the *Growing* page so they can record the growth of their plants every couple of weeks, depending on how fast the plants grow.
5. On the *Observing* page of the journal, the students can record their observations by either drawing pictures of the plant or writing about what they see.

PRIMARILY PLANTS

6. Choose an area in the room where students can place the seeds they germinate and grow. This area should have easy access but be somewhat protected so that the seeds aren't harmed during daily activities. Once seeds have sprouted it will be necessary to have them in a location where they can get sunlight.

Procedure

Germination of Sunflower Seeds
1. Ask the students how many have seen a sunflower plant. Invite them to describe it.
2. Ask what they know about the growth of a sunflower seed. Wait for responses, and if desired, write the responses on chart paper. Tell the students they will be observing the growth of a sunflower plant starting from a seed.
3. Give each group of students four sunflower seeds. Provide each group with a zipper-type plastic bag, and a paper towel. Tell the students to fold the paper towel so it will fit into the plastic bag. Have them dampen the paper towel by spraying it with water. Tell them to place the seeds on the damp towel and then to slide the towel with the seeds into the plastic bag and seal it. Direct the students to put their seed bags in the designated area.
4. Ask the students why they think they were told to add water to the paper towel. Explain to the students that a seed needs to absorb water until it breaks its seed coat and the tiny embryo starts to grow sending out a root and a stem. Tell them they will watch a sunflower seed germinate—begin to grow.
5. Invite the students to observe the seeds carefully and record the growth of the seeds in their journals by drawing and writing about them as they germinate.

Planting the Seeds
1. Distribute the seed bags, plastic cups, soil, and permanent markers. Have students use the marker to write their group's name on the cup. Instruct the students to fill cups with moistened soil. Then have them gently pull the paper towel and seeds out of the bag.
2. Show them how to carefully plant the sprouted seeds by placing one seed between the soil and side of the cup. This will allow the students to watch the growth of the plant's root below the soil line. Caution them to be very careful to not break off the roots or stems of the plants. If any accidents do occur, replace the damaged seed with a new seed. Discuss how this new seed will need to go through the germination process.
3. Encourage students to help each other as they plant their seeds in the cups. Ask a few students to plant extra seeds in some cups for the class or school garden. Instruct the students to record in their journals how they planted the sunflower seeds by drawing their seeds on the page *Planting*.
4. Place the cups in a sunny spot. Suggest that the students cover their cups with thin plastic wrap so that watering will not be necessary until the small sunflower plants start to grow above the soil line.
5. In about a week, uncover the cups. Caution the students to water the plants enough so they will grow but not to overwater them.
6. Provide time for the students to measure the plants' growth for about a week or until all the seedlings are growing well.
7. As the plant grows, encourage the students to predict how many days it will take for the leaves to begin to grow. Instruct the students to report in their journal the actual results and compare their observations with their classmates.
8. Have the students take their seedlings home. As a class, transplant the extras outdoors if the weather will permit their continued growth.

Observing the Sunflower Plant
1. After the sunflower plant has been transplanted outdoors and is growing well, tell the students to measure the height of the plant weekly and record those measurements on the page *Growing* in their journals.
2. Ask them how many leaves they think are on the plant. Ask them if all the leaves are the same size. Have them observe and discuss where the smaller leaves are growing. Ask why they think the small leaves are located where they are.
3. What does the leaf of the plant feel like?
4. Invite the students to observe and describe the stem of the plant. Tell them to measure its height and circumference.
5. As the plant starts to bud, ask the students to describe the shape of the unopened sunflower.
6. Question the students as to which way the young plant faces in the morning. Is it different from the way it is turned in the afternoon?
7. As the sunflower plant grows, have the students predict how many days it will take for their plants to flower. Have them record their predictions in their journal.
8. Encourage the students to watch their sunflower plants and record when the plant produces a flower, when the flower opens, and when it produces seeds. Invite them to record this information in their journal on the page *Flowering*.

Connecting Learning
1. What does germinate mean?
2. How many days did it take for the first seed to sprout? Why didn't the seeds all sprout at the same time?
3. Describe the germination of your seed. What part of the new plant is first to appear from the seed? What comes next?
4. Why do you think it is important to the plant that the root grows first?
5. What helps the seed to grow?
6. How long did it take for the first leaves to appear on the plant? What do the leaves do for the new plant?
7. Why do you think some seeds grew faster (or slower) than others?

Home Link
Copy the sheet showing the growth cycle of a plant. Send it home with the students. Encourage the students to cut the picture cards apart, put them in the correct order, and explain to their parents how their sunflower seed grew.

Evidence of Learning
Students should be able to explain how their seeds germinated. They should explain after the seed absorbed water, the root emerged first, then the stem and leaves.

Key Question

How does a sunflower seed grow into a mature plant?

Learning Goals

Students will:

- germinate sunflower seeds, and
- plant the seeds and observe the growth cycle.

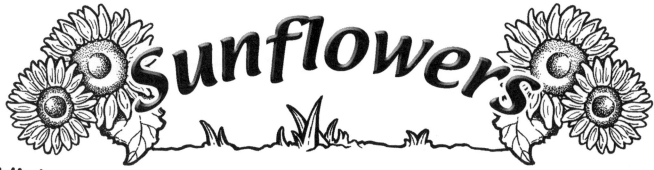

History

The wild sunflower (Helianthus Annuus) is a native of North America and one of our most familiar garden plants. The American Indians in the United States have been using and cultivating the sunflower for thousands of years. Evidence suggests that they began cultivating and improving the sunflower in the Four Corners area of southwestern United States as early as 3000 B.C. It was the American Indian who first domesticated the plant into a single-headed plant with a variety of seed colors. They used the wild sunflower seeds for food and medicine. The seeds were usually roasted and ground into a meal for baking or used to thicken soups and stews. Roasted sunflower hulls were used to make a coffee-like beverage. Yellow dye, which was obtained from the flowers, and a black or dull blue dye, obtained from the seeds, were important in Indian basketry and weaving. Oil, extracted from ground-up seeds by boiling, was used for cooking oil and hair treatment. The dried stalk was used as building material.

When the early explorers came to the New World, they found that the Aztecs revered the sunflower and used it in their temples of the Sun. The sunflowers were represented in gold in these temples.

Explorers took the sunflower plant to Europe in the 1500s where it was used as a curiosity and an ornamental plant. The Russians were the first to make commercial use of the plant by manufacturing sunflower oil. By the early 1800s, Russian farmers were growing over two million acres of sunflowers. By the late 1800s, the improved Russian sunflower seed found its way back into the United States, probably by some immigrants.

Growth

The sunflower is a tall plant with bright yellow flowers. It is an annual plant that has a rough hairy stem and grows three to 10 feet tall. The rough heart-shaped leaves are three to 12 inches long. The flower heads are composed of a disk of many small tubular flowers arranged compactly in a swirl and are surrounded by a fringe of large yellow petals that forms the rays of the composite flower. The head can be as large as 12 inches in diameter and produce up to 1000 seeds. The plant flowers from July to October.

The head of the sunflower follows the sun each day. In the morning it faces the rising sun in the east. It follows the westward movement of the sun, so that at sunset it faces west. Sunflower in Spanish means "looks at the sun."

Sunflowers are easy to grow as long as they have sunny locations. Good fertile soil will provide for large flower heads and meaty seeds. Planting usually begins in early May with germination occurring in seven to 12 days. Plants will mature and start to flower in 80 to 90 days.

Harvest begins by the middle of September and continues into October. When the head shrivels and turns down and the seeds are ripe, the plants are cut at ground level and stood with the head uppermost like corn stalks. When dry, the seeds are removed by hitting the backsides of the heads with a mallet. The seeds are then stored in bags in a dry place. If there are just a few plants, the head can be cut with about a foot of stem attached and hung in a warm dry place. A bag can be placed over the heads to catch falling seeds as they drop during drying. Once the seeds are dried, they can be rubbed easily from the seed heads.

Uses

Sunflowers are more than just pretty plants; the seeds are a rich treasure of vitamins, minerals, protein, polyunsaturated fat, and fiber. Sunflower seeds contain a good amount of vitamin E, B complex, minerals, such as magnesium, iron, potassium, and calcium. There is no cholesterol! However sunflower seeds have a lot of calories, one-half cup of hulled seeds contains about 400 calories.

Economically every part of the sunflower can be used for some purpose. The leaves are used for cattle feed, the stems contain a fiber that can be used in making paper, and oil is obtained from the seeds.

Sunflower seeds provide a high-quality vegetable oil. To extract the oil, the seeds are crushed and ground to meal. The oil pressed from the seeds is of a yellow color and considered equal to olive oil or almond oil for cooking and table purposes. The residue meal cake left after the oil is pressed out forms a valuable food for cattle, sheep, pigs, and poultry.

The sunflower is one of only four major crops of world-wide importance that is native to the United States. (The other three are the cranberry, pecan, and blueberry.) There are several million acres devoted to growing sunflowers in the United States.

Planting

This is how I planted my sunflower.

I predict _____ days for leaves to grow.

I count _____ days for leaves to grow.

2.

✂

1.

The Story of a Sunflower

How does a sunflower seed grow into a mature plant?

PRIMARILY PLANTS 144 © 2005 AIMS Education Foundation

Growing

The sunflower is growing.

Week: _____

Height: _____

Number of leaves: _____

Size of head: _____

Picture

4.

Germinating

I put the seeds on the damp towel and close the bag.

☐ I predict _____ days to germinate.

☐ I count _____ days to germinate.

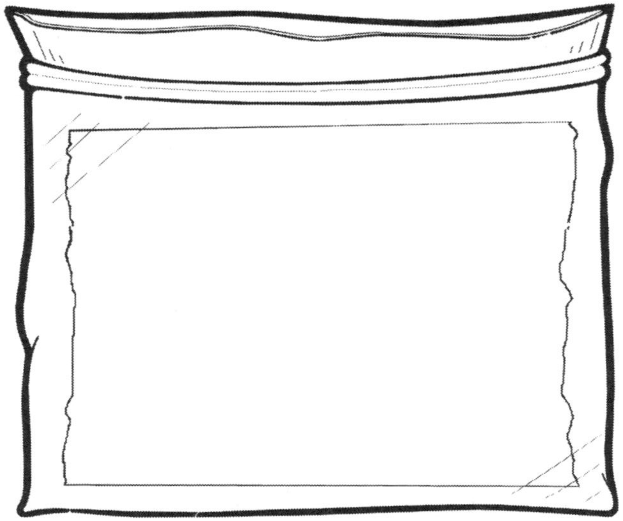

My seeds grew _____

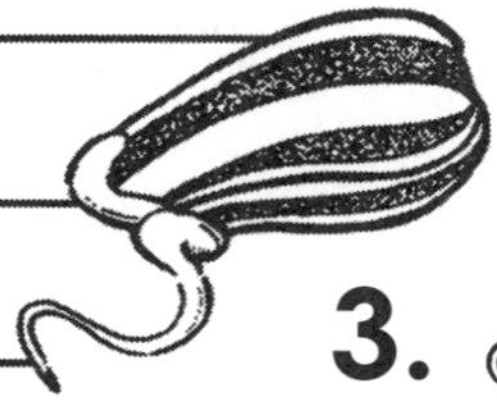

3.

Observing

I observed sunflower seeds grow into a mature plant.

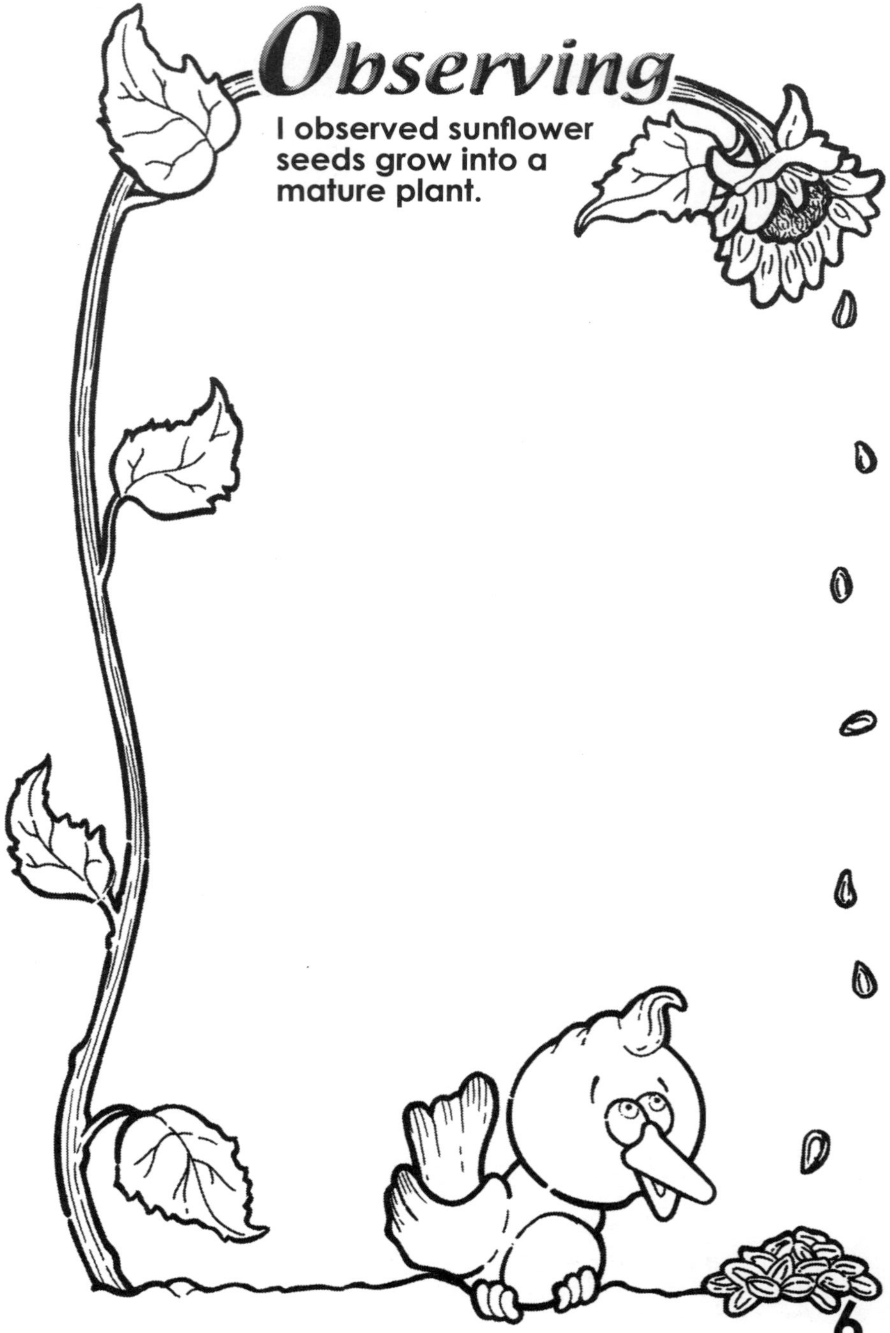

6.

Flowering

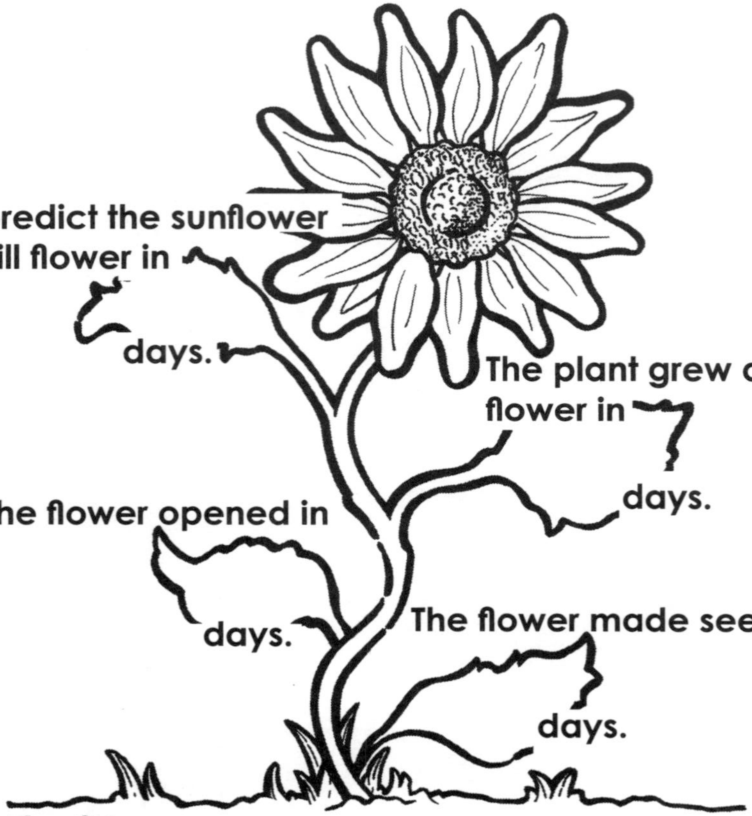

I predict the sunflower will flower in _____ days.

The flower opened in _____ days.

The plant grew a flower in _____ days.

The flower made seeds in _____ days.

The flower…

_____
_____
_____
_____

5.

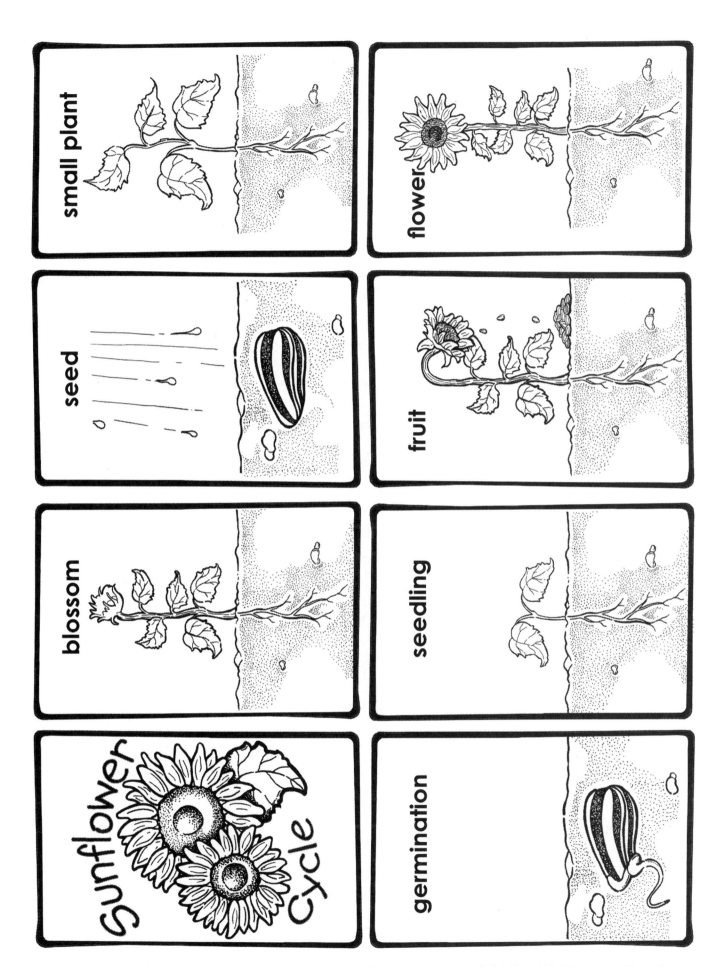

Connecting Learning

1. What does germinate mean?

2. How many days did it take for the first seed to sprout? Why didn't the seeds all sprout at the same time?

3. Describe the germination of your seed. What part of the new plant is first to appear from the seed? What comes next?

4. Why do you think it is important to the plant that the root grows first?

5. What helps the seed to grow?

6. How long did it take for the first leaves to appear on the plant? What do the leaves do for the new plant?

7. Why do you think some seeds grew faster (or slower) than others?

People Need Plants

Topic
Plant uses

Key Question
Why are plants important to people?

Learning Goal
Students will learn that many things they use in everyday life come from plants.

Guiding Documents
Project 2061 Benchmark
- *Animals eat plants or other animals for food and may also use plants (or even other animals) for shelter and nesting.*

NRC Standards
- *Objects are made of one or more materials, such as paper, wood, and metal. Objects can be described by the properties of the materials from which they are made, and those properties can be used to separate or sort a group of objects or materials.*
- *All animals depend on plants. Some animals eat plants for food. Other animals eat animals that eat the plants.*

Science
Life science
  botany
    plants

Integrated Processes
Observing
Collecting and recording data
Organizing

Materials
Materials made from plants (see *Management 1*)
Pictures of plant products (see *Management 2*)
Pictures of trees, cotton plants, and corn plants
Student page

Background Information
Plants are very important to people in many ways. When plants make their food, they take in carbon dioxide and give off oxygen, which is essential to all animal life on Earth. All of the food we eat comes from plants, either directly or indirectly. In addition to our physical needs, we use plants for many other things including shelter, medicine, and clothing.

Management
1. Collect as many examples of things that come from plants as possible. Some possibilities include vegetables, wood, flowers, paper, cotton fabric, seeds, etc.
2. Supplement your collection of real items with pictures of other ways plant products are used such as in construction, clothing, furniture, and body care products.
3. Before doing the activity, make an area on a bulletin board or wall with the following labels: *Homes, Clothing, Food, Body Care Products,* and *Things We Use.*

Procedure
1. Show students the pictures of trees, cotton plants, and corn plants. Discuss how humans use these plants.
2. Give students the pictures and other examples you collected of plants and plant products. Ask them to put the items under the correct headings on the wall.
3. Do a survey to find out how many students are wearing cotton clothing. Have the class make a real graph comparing those wearing cotton to those wearing other fabrics.
4. Have students each find at least one thing in the room that comes from a plant. Provide a time of sharing where everyone tells what he or she found and what plant or plant part it comes from.
5. Ask students to share their favorite breakfast meals. Discuss what plants those foods come from.
6. Distribute the student page and help students make a list of all the things that would be missing from the room if there were no plants.

Connecting Learning
1. What are some of ways that people use plants?
2. What would happen if there were no plants? Why?
3. What parts of your lunch come directly from plants? [fruit, vegetables, bread, peanut butter, jam, etc.] What parts come indirectly from plants? [lunchmeat, milk, etc.]
4. What are you wondering now?

Extensions
1. Make a collage of things that come from plants—seeds, leaves, stems, flowers, etc.
2. Have students write a thank-you letter to a plant.

PRIMARILY PLANTS © 2005 AIMS Education Foundation

People Need Plants

Key Question

Why are plants important to people?

Learning Goal

Students will:

learn that many of the things they use in everyday life come from plants.

Connecting Learning

1. What are some of ways that people use plants?

2. What would happen if there were no plants? Why?

3. What parts of your lunch come directly from plants? What parts come indirectly from plants?

4. What are you wondering now?

Plant Parts

Each part of a plant has an important function.

Flowers are the reproductive parts of a plant. Flower petals and the flower's smell attract bees and insects to pollinate the flower. After pollination, the petals fall away and seeds develop in the part of the flower called the ovary. The ovary itself usually becomes what we call fruit.

Stems support the upper parts of plants Water and dissolved nutrients from the soil travel up the stem in a system of tubes. Food from the leaves travels down the stems to the roots. Stems also store food.

Leaves are the parts of the plant where food is made by photosynthesis. Leaves take in carbon dioxide from the air, water from the soil, and energy from sunlight. During photosynthesis, the leaves use light energy to change carbon dioxide and water into sugars (food).

Roots of plants anchor the plants in the soil. Water and minerals are taken from the soil through the roots. Many plants, such as carrots, store food in their roots.

Seeds contain a tiny embryo of a plant inside. The seed contains food that supplies energy and materials for growth until the plant grows its first leaves above the ground.

Observe A Leaf

Topic
Leaves

Key Question
What does a leaf look like?

Learning Goal
Students will observe and describe leaves.

Guiding Documents
Project 2061 Benchmark
- Tools such as thermometers, magnifiers, rulers, or balances often give more information about things than can be obtained by just observing things without their help.

NRC Standards
- Each plant or animal has different structures that serve different functions in growth, survival, and reproduction. For example, humans have distinct body structures for walking, holding, seeing, and talking.
- Ask a question about objects, organisms, and events in the environment.

*NCTM Standard 2000**
- Use tools to measure

Math
Measurement
    linear

Science
Life science
    botany
        leaves

Integrated Processes
Observing
Comparing and contrasting
Collecting and recording data

Materials
Leaves, one per student
Hand lenses
Crayons or colored pencils
Student page

Background Information
The variety found among the leaves of plants is enormous. There are large leaves, small leaves, slender leaves, and wide ones. Leaves can be soft, prickly, hairy, and hard. But leaves all have one thing in common—they change sunlight into energy through photosynthesis. The leaves absorb carbon dioxide from the air, and, with water that comes through the roots of the plant, combine these elements and release oxygen into the air. By this exchange, plants maintain a level of oxygen in the air that benefits all living things.

Management
1. Ideally students will select a leaf to study while out on a nature walk. If this is not possible, have them bring leaves to study from home.
2. The leaves should be fresh; they will be hard for the students to work with if they dry out and get brittle.

Procedure
1. Take students on a nature walk to select their leaves (or have them bring their leaves from home).
2. Tell students to take their special leaves and look at them carefully.
3. Distribute the student page and crayons or colored pencils. Ask students to carefully draw their leaves in the space provided on the page.
4. Distribute the hand lenses and have students look at the veins on their leaves.
5. Have them use the ruler on the student page to measure their leaves and record the length and width.
6. Instruct them to complete the page by filling in their observations.
7. On the back of the student page, have them trace their leaves and use the crayons to color them appropriately.

Connecting Learning
1. Describe your leaf.
2. How is your leaf like the leaves of your classmates? How is it different?
3. Do all leaves look alike? How do you know?
4. Are all leaves green? How do you know?
5. What do leaves do for a plant?
6. What are you wondering now?

Extensions
1. Make a collection of as many different kinds of leaves as you can find.
2. Sort and classify the leaves you have collected in different ways.
3. Make leaf prints or leaf rubbings using the leaves collected.
4. Glue the leaves between two pieces of waxed paper and make a construction paper frame. Display them in the window.

* Reprinted with permission from *Principles and Standards for School Mathematics*, 2000 by the National Council of Teachers of Mathematics. All rights reserved.

Observe A Leaf

Key Question

What does a leaf look like?

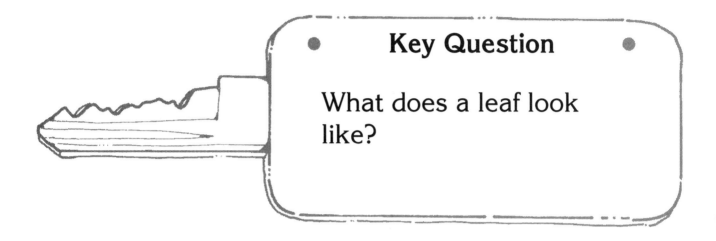

Learning Goal

observe and describe leaves.

Observe A Leaf

Here is a drawing of my leaf.

1. I found my leaf _____.

2. My leaf is _____ cm long and _____ cm wide.

3. My leaf is colored _____.

4. My leaf smells like _____.

5. My leaf feels like _____.

6. My leaf is important to its plant because

_____

_____.

Connecting Learning

1. Describe your leaf.

2. How is your leaf like the leaves of your classmates? How is it different?

3. Do all leaves look alike? How do you know?

4. Are all leaves green? How do you know?

5. What do leaves do for a plant?

6. What are you wondering now?

Leafy Facts

Simple Leaf

The broad, flat part of a leaf is called the blade. The blade is connected to the stem or petiole. The petiole supports the blade and turns it toward the sun. The bud grows at the base of the leaf. The midrib is the central stalk of the leaf.

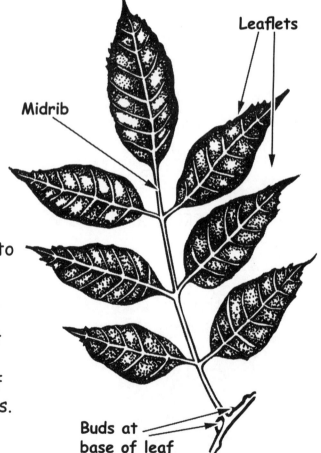

Compound Leaf

In some plants, the blades are divided into a number of small leaves called leaflets. A compound leaf has a number of leaves arranged in two rows facing each other along the midrib. One leaflet may grow at the tip of the midrib of some compound leaves. Buds always appear at the base of the leaf, never at the base of the leaflets.

PRIMARILY PLANTS © 2005 AIMS Education Foundation

More Leafy Facts

A blade of grass, a pine needle, a fern frond, and a maple leaf are all leaves. Most leaves need light, air, and water. Leaves contain green cells to make food for the entire plant.

Leaves look very different, however, Every plant has its own distinctive kind of leaf. Leaves help us identify plants like fingerprints identify people.

Needle-like leaves can be divided into two groups. Needles can be attached singly to a stem. Other needle-like leaves have needles attached in bundles of two, three, and four.

Almost all other leaves can be grouped into broad leaves. Broad leaves can be further classified by their pattern of veins.

In palms, grasses, and other plants, the veins run parallel to one another from the petiole to blade tip.

The branched, or webbed group, can be classified into two groups—pinnate (featherlike) or palmate (fanlike).

In the pinnate group, the veins branch out from the midrib like barbs of a feather.

The veins in a palmate leaf fan out from the petiole and form a network of smaller veins through the leaf.

Simple Leaf

Compound Leaf

Broad Leaf

Needle-like Leaf

Pinnate Veins

Palmate Veins

PRIMARILY PLANTS © 2005 AIMS Education Foundation

Leaf Safari

Topic
Leaves

Key Question
What are our leaves like?

Learning Goals
Students will:
- compare, measure, and describe leaves;
- sort leaves according to various attributes; and
- learn the purpose of leaves and veins.

Guiding Documents
Project 2061 Benchmarks
- Tools such as thermometers, magnifiers, rulers, or balances often give more information about things than can be obtained by just observing things without their help.
- Most living things need water, food, and air.

NRC Standards
- Each plant or animal has different structures that serve different functions in growth, survival, and reproduction. For example, humans have distinct body structures for walking, holding, seeing, and talking.
- Ask a question about objects, organisms, and events in the environment.

NCTM Standard 2000*
- Use tools to measure
- Sort, classify, and order objects by size, number, and other properties
- Sort and classify objects according to their attributes and organize data about the objects

Math
Measurement
　linear
Sorting

Science
Life science
　botany
　　leaves

Integrated Processes
Observing
Comparing and contrasting
Collecting and recording data

Materials
Leaves (see *Management 1*)
Leaf-collecting bag, one per student
Hand lenses
Crayons or colored pencils
Grouping circles (see *Management 3*)
Rulers
Long paper (see *Management 4*)
Student pages

Background Information
Green plants are the only living things that make their own food. Scientists call this process *photosynthesis*, which means, "making use of light." Green leaves contain a substance called chlorophyll that turns the light energy from the sun into food energy for the plant. Leaves take carbon dioxide from the air and water and convert it into glucose, oxygen, and water. The glucose (simple sugar) is carried to other parts of the plant through the veins in the leaves. The oxygen is released into the air through tiny pores or openings in the leaf called stomata. Leaves need sunlight in order to retain chlorophyll and produce food. All animal life on Earth depends on plants for food.

There is great variety in the shapes, margins, surfaces, and textures of leaves. A leaf is simple when it has one blade on one leaf stalk. It is compound when the blade is composed of several separate leaflets on one leaf stalk. Leaves may be long and narrow with the veins running parallel to the edges. In others, the veins branch out like fans from the bases of the leaves. These are called palmate leaves. In pinnate leaves, veins are borne on the midrib and branch out along its length. The margins, or edges, of leaves may be smooth, have sharp teeth, or be deeply cut into fairly large portions, called lobes. The surface and texture of a leaf may be smooth, rough, or harsh to the touch.

PRIMARILY PLANTS　　　　© 2005 AIMS Education Foundation

Management
1. It is important for the class to go outside and collect leaves for the leaf safari. Each student should be encouraged to find as many different kinds of leaves as possible.
2. Brown paper lunch sacks work well for the leaf-collecting bags.
3. If grouping circles are not available, you may substitute large loops of colored yarn.
4. Students will need long pieces of paper on which to make leaf rubbings of five to six leaves.

Procedure
1. Give each student a leaf-collecting bag. Take the class outside on a guided leaf safari. Tell them to collect a variety of leaves.
2. Return to the classroom and distribute the first student page. Have students complete the page.
3. Discuss the similarities and differences among the leaves collected. Use grouping circles to group the leaves into sets. Groupings could include long and thin, broad and flat, needlelike, jagged edges, shiny, dull, rough, smooth, etc.
4. Choose two attributes (large and shiny, for example). Make a two-circle Venn diagram using the grouping circles and place all the leaves in the proper set. Describe each set by attribute and number in the set.
5. Distribute hand lenses and have students observe their leaves. Have them identify the veins and notice their shapes. Discuss the idea that these veins carry food, water, and nutrients to all parts of the leaf. Compare that idea to the veins in their own hands and arms and the similar function of carrying blood and nutrients to parts of the body.
6. Give students rulers and have them measure the lengths of five or six different leaves. Instruct them to sequence those leaves from shortest to longest and make a leaf rubbing on a long piece of paper to record the order.
7. Have the students each select one leaf from their set. Distribute the second student page and instruct them to carefully trace the leaf on the grid. Have them count the lengths, widths, and surface areas of their leaves.
8. As a class, order the leaves from least surface area to greatest surface area.

Connecting Learning
1. What kinds of leaves did you find on the leaf safari?
2. Describe the similarities and differences among the leaves.
3. In what ways were you able to sort and classify the leaves?
4. What did you notice when you observed your leaves with a hand lens?
5. What is the purpose of a leaf? [to make food for the plant]
6. What is the purpose of the veins in a leaf? [to move the food throughout the leaf] How is this similar to the purpose of the veins in our bodies?
7. What are you wondering now?

Extensions
1. Study the types of leaves that we eat for food. Prepare vegetables for students to taste—both raw and cooked. Let the students vote on which vegetable leaves they like and how they like them prepared. Edible leaves include kale, spinach, lettuce, parsley, collards, mustard greens, and brussel sprouts.
2. Decorate T-shirts with leaf rubbings.
3. Make a collection of leaves by drying them and mounting them on a stiff piece of cardboard.

* Reprinted with permission from *Principles and Standards for School Mathematics*, 2000 by the National Council of Teachers of Mathematics. All rights reserved.

Leaf Safari

Key Question

What are our leaves like?

Learning Goals

Students will:

- compare, measure, and describe leaves;
- sort leaves according to various attributes; and
- learn the purpose of leaves and veins.

Leaf Safari

I picked up _____ leaves outside.

Draw a picture of your leaf collection in the bag.

Write some words that describe your leaves.

1. Pick your favorite leaf.
2. Describe the color and shape.

3. Trace your leaf on the grid.

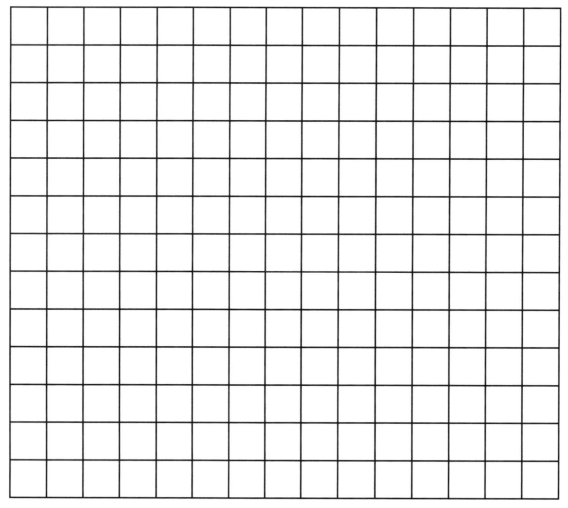

4. Measure, count, and record:

   Leaf length:_____   Leaf width:_____

   Leaf area:_____ squares

Connecting Learning

1. What kinds of leaves did you find on the leaf safari?

2. Describe the similarities and differences among the leaves.

3. In what ways were you able to sort and classify the leaves?

4. What did you notice when you observed your leaves with a hand lens?

5. What is the purpose of a leaf?

6. What is the purpose of the veins in a leaf? How is this similar to the purpose of the veins in our bodies?

7. What are you wondering now?

Stem Study

Topic
Stems

Key Question
What are the functions of a plant stem?

Learning Goal
Students will learn how stems are necessary to plants.

Guiding Documents
Project 2061 Benchmark
- Plants and animals have features that help them live in different environments.

NRC Standards
- Each plant or animal has different structures that serve different functions in growth, survival, and reproduction. For example, humans have distinct body structures for walking, holding, seeing, and talking.
- Ask a question about objects, organisms, and events in the environment.

Science
Life science
  botany
    stems

Integrated Processes
Observing
Predicting
Comparing and contrasting
Collecting and recording data

Materials
Various plant stems (see *Management 1*)
Celery stalks
White daisy or carnation
Food coloring
Plastic cups
Student pages

Background Information
The stems of plants serve many functions. One function is to support the plant parts that are above the ground. The stem holds up the plant's parts toward the sun so the plant can receive the light energy it needs. Some plants have stems that are soft and green. Others have stems that are thick and hard, like trees.

The most important function of a stem is to serve as a transport system in plants. Small tubes from the roots go up through the stems. Water and minerals are carried from the roots to the leaves of a plant. Food made in the leaves moves through the tubes in the stem to other parts of the plants. Some stems are specialized organs used to store food. Stem plants that we eat include celery, asparagus, sugar cane, broccoli, and potatoes.

Management
1. Collect a variety of stems that are familiar to students as food they eat, such as celery, broccoli, asparagus, rhubarb, and potatoes.
2. You will need enough celery for each student to have one stalk.
3. For the "Flower in Water" experiment, you will need one large white daisy or carnation.

Procedure
Day One
1. Take the students outside to identify plants and look at the stems of plants. Compare and contrast the different plant stems you see. Tree trunks are hard, thick stems. Many flowers have soft, thin stems.
2. Discuss the functions of stems (see *Background Information*).
3. Have the class try to identify some stems that we eat. [celery, broccoli, asparagus, rhubarb, sugar cane, potatoes] Show them the examples of edible stems that you have. Explain that potatoes are actually underground stems, not roots.
4. Give each student a stalk of celery and the first student page. Have them draw their celery stems in the space provided. Ask them to describe their celery stems. List their descriptions on the board.
5. Tell students what they will be doing with the celery, and ask them to predict what will happen.

6. Distribute cups of water and food coloring and have students put their stems in the water and leave them overnight. Use different colors of water for variety.
7. Set up the "Flower in Water" experiment and have students share their predictions of what will happen.

Day Two
1. Have students break open their stalks of celery to see the inside tubes. Ask them to record the results on the student page.
2. Distribute the "Flower in Water" page to each student and have them draw in the results by coloring the flower.
3. Discuss students' predictions and the actual results.

Connecting Learning
1. What is the purpose of a stem?
2. How are stems alike? How are they different?
3. Do all stems grow above ground?
4. What happened when you put the celery in colored water and left it overnight? Is this what you predicted? Why or why not?
5. What happened when you put the daisy in colored water and left it overnight? Is this what you predicted? Why or why not?
6. What does this tell you about how food and water travel through stems and into leaves and flowers?
7. What are you wondering now?

Extensions
1. As an extension to the "Flower in Water," do the "Colorful Changes." Fill three glasses with water. Color one blue, one red, and in the third, mix red and blue to make purple. Put one white flower (daisy or carnation) in each glass. Have the students predict what will happen to each one. Most students will predict that the flowers will turn red, blue, and purple. What actually happens is sometimes surprising. The blue color will rise, the red color will rise, but the mixed purple often will separate back into red and blue as it rises.
2. Make rubbings of various kinds of stems.
3. Discuss the various uses of tree stems.

Stem Study

Key Question

What are the functions of a plant stem?

Learning Goal

lean how stems are necessary to plants.

Stem Study

Here is a drawing of my stem.

My stem is from a

1. Put some water in a glass.
2. Add four drops of food coloring.
3. Cut off the end of the stem.
4. Place the stem in the glass.
5. Leave the stem overnight.

What happened?

Why?

What are stems for?

Stem Study
Flower in Water

Do this:
1. Pour water in a glass.
2. Color the water red.
3. Put a flower in the water.
4. Leave it overnight.

5. How did the flower change?

Stem Study
Colorful Changes

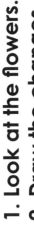

1. Add food coloring to the cups.
2. Put a flower with stem in each cup.
3. Draw the three flowers. Color them.
4. Leave the flowers overnight.

1. Look at the flowers.
2. Draw the changes.

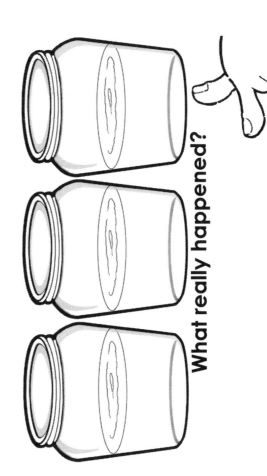

What really happened?

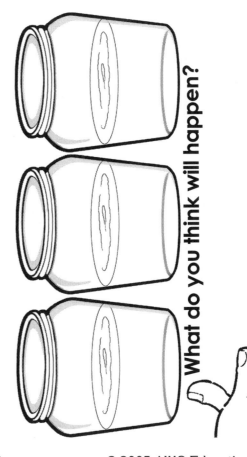

What do you think will happen?

Connecting Learning

1. What is the purpose of a stem?

2. How are stems alike? How are they different?

3. Do all stems grow above ground?

4. What happened when you put the celery in colored water and left it overnight? Is this what you predicted? Why or why not?

5. What happened when you put the daisy in colored water and left it overnight? Is this what you predicted? Why or why not?

6. What does this tell you about how food and water travel through stems and into leaves and flowers?

7. What are you wondering now?

Super Tuber

Topic
Tubers

Key Question
How is a potato different from other kinds of plant stems?

Learning Goal
Students will learn that potatoes are underground stems.

Guiding Documents
Project 2061 Benchmarks
- Plants and animals have features that help them live in different environments.
- Tools such as thermometers, magnifiers, rulers, or balances often give more information about things than can be obtained by just observing things without their help.

NRC Standards
- Each plant or animal has different structures that serve different functions in growth, survival, and reproduction. For example, humans have distinct body structures for walking, holding, seeing, and talking.
- Ask a question about objects, organisms, and events in the environment.
- Employ simple equipment and tools to gather data and extend the senses.

NCTM Standards 2000*
- Use tools to measure
- Represent data using concrete objects, pictures, and graphs

Math
Measurement
  mass
  length
Graphing
Venn diagrams

Science
Life science
  botany
    stems

Integrated Processes
Observing
Comparing and contrasting
Collecting and recording data
Interpreting data

Materials
Sack of potatoes
Crayons or colored pencils
Metric rulers
Balances
Teddy Bear Counters
Sticky notes in two colors
Student pages

Background Information
The potato, though it grows underground, is not a root of the plant, but the stem. The potato plant has stems, roots, leaves, and flowers. There are swellings on the underground parts of the stems called "tubers." These tubers are what we call potatoes.

Look carefully at a potato. The "eyes" are tiny buds with a small scale-like leaf beside each eye. If you cut an eye from a potato and plant it in the soil, the bud will grow a new plant.

Potatoes are a basic food for millions of people throughout the world. The potato was first grown in South America and spread from there to Europe and North America by Spanish travelers.

Management
1. Make sure the sack of potatoes contains enough for each student to have one.
2. Use balances and Teddy Bear Counters for students to find the mass of their potatoes.

Procedure
1. Bring in a sack of potatoes hidden in a grocery bag. Pass the bag around and let students feel the bag and guess what they think is inside.
2. Play a quick game of 20 questions to allow the students to determine what is in the bag. Tell students they must ask questions that can be answered with "yes" or "no." When someone guesses correctly, bring the sack of potatoes out of the bag to show the class.
3. Explain that the potato is actually an underground stem, not a root (see *Background Information*).

PRIMARILY PLANTS © 2005 AIMS Education Foundation

4. Give each student a potato to study. Distribute the first student page and have them draw pictures of their tubers.
5. Make the measuring tools available and give everyone time to find the length and mass of his or her potato.
6. On the board or an overhead, make a bar graph of students' potato preferences. Distribute the graphing page and have them record the information on their own pages.
7. Make a multi-circle Venn diagram on the board with at least three of the potato preferences. Have students put a sticky note with their name on it in the portion of the Venn that reflects their preferences. (Be sure to point out that unlike the bar graph, the Venn diagram allows them to show a preference for more than one kind of potato.) Use different color notes for boys and girls.

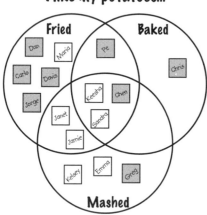

Connecting Learning
1. Describe your potato. How many eyes does it have? What do the eyes look like? What is its mass? How long is it?
2. Are potatoes roots? [no] What are they? [part of the underground stem]
3. Where are the other parts of the potato plant (roots, leaves, flowers)?
4. What is the most popular way to eat potatoes in our class? How do you know?
5. How is the Venn diagram like the bar graph? [Both show class preferences for potatoes.] How is it different? [The Venn diagram allows people to have a preference for more than one kind of potato.]
6. According to the Venn diagram, do more girls or boys like French fries? ...mashed potatoes? ...baked potatoes? ...etc.?
7. What are you wondering now?

Extensions
1. Make potato prints.
2. Leave a potato in a dark closet for several weeks. Observe the changes.
3. Plant an eye from a potato and see how it grows.

* Reprinted with permission from *Principles and Standards for School Mathematics*, 2000 by the National Council of Teachers of Mathematics. All rights reserved.

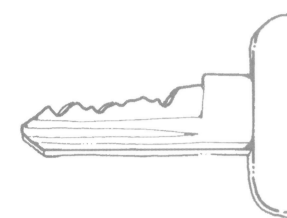

Key Question

How is a potato different from other kinds of plant stems?

Learning Goal

learn that potatoes are underground stems.

1. My tuber is called a _____.

2. Here is a picture of my tuber.

Count and Measure

3. My pototo has _____ eyes.

4. My potato is _____ cm long.

5. The mass of my potato is _____ teddy bears.

6. I like to eat my potatoes _____.

PRIMARILY PLANTS

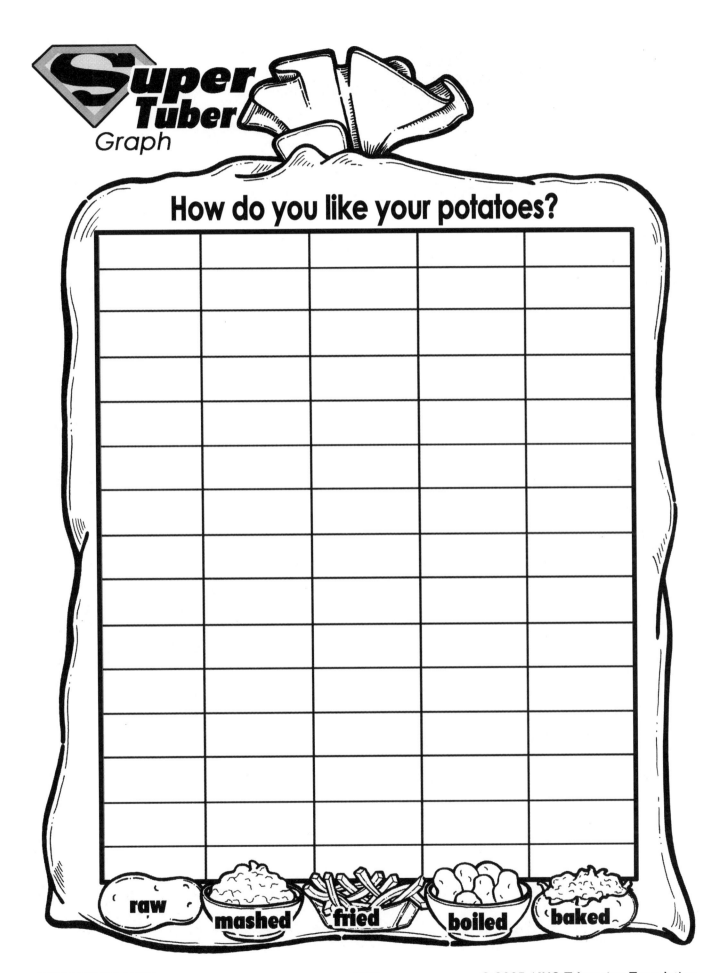

Connecting Learning

1. Describe your potato. How many eyes does it have? What do the eyes look like? What is its mass? How long is it?

2. Are potatoes roots? What are they?

3. Where are the other parts of the potato plant (roots, leaves, flowers)?

4. What is the most popular way to eat potatoes in our class? How do you know?

5. How is the Venn diagram like the bar graph? How is it different?

6. According to the Venn diagram, do more girls or boys like French fries? ...mashed potatoes? ...baked potatoes? ...etc.?

7. What are you wondering now?

Root Study

Topic
Roots

Key Questions
1. Why do plants need roots?
2. Do all root systems look alike?

Learning Goals
Students will:
- learn the functions of plant roots, and
- learn about two kinds of root systems—taproots and fibrous roots.

Guiding Documents
Project 2061 Benchmarks
- *Plants and animals have features that help them live in different environments.*
- *Tools such as thermometers, magnifiers, rulers, or balances often give more information about things than can be obtained by just observing things without their help.*

NRC Standards
- *Each plant or animal has different structures that serve different functions in growth, survival, and reproduction. For example, humans have distinct body structures for walking, holding, seeing, and talking.*
- *Ask a question about objects, organisms, and events in the environment.*
- *Employ simple equipment and tools to gather data and extend the senses.*

*NCTM Standard 2000**
- *Use tools to measure*

Math
Measurement
  length

Science
Life science
  botany
    roots

Integrated Processes
Observing
Comparing and contrasting
Collecting and recording data

Materials
Root samples (see *Management 1*)
Hand lenses
Student page

Background Information
Plant root systems are essential to plants. The function of the roots is to anchor plants and absorb water and nutrients from the surrounding soil. There are different kinds of root systems. Taproots are large central roots that grow deep in the soil. Smaller roots extend from this central root. A carrot is an example of a taproot. Fibrous roots are shallow and spread over a wide area. They branch considerably and have a threadlike appearance. Grass has a fibrous root system.

Management
1. Collect several examples of fibrous roots and taproots. Suggestions include grass, dandelions, clover, radishes, and carrots. Wash the soil from the roots before giving them to students to observe.

Procedure
1. Ask the *Key Questions* and have students share their thoughts.
2. Show students the plants and roots you have collected. Distribute the plants and hand lenses and allow students time to observe the roots.
3. Introduce the terms *taproot* and *fibrous root*. Describe the differences between these kinds of roots (see *Background Information*). Explain that the roots absorb water and nutrients to feed the plant. They also help anchor it in the ground.
4. Distribute the student page and have each student select one root to draw and describe.

Connecting Learning
1. Why do plants need roots?
2. Do all roots look alike? How are they the same? How are they different?
3. Would a tree have a taproot or fibrous roots? Why do you think so?
4. What kind of roots do many weeds have? Why might this be?
5. What are you wondering now?

Extensions
1. Take a cutting from a philodendron, Wandering Jew, Creeping Charlie, or similar plant and put it in water. Watch as the roots develop.
2. Research the kinds of roots that some of students' favorite flowers and trees have.

* Reprinted with permission from *Principles and Standards for School Mathematics*, 2000 by the National Council of Teachers of Mathematics. All rights reserved.

Root Study

Key Questions

1. Why do plants need roots?
2. Do all root systems look alike?

Learning Goals

Students will:

- learn the functions of plant roots, and
- learn about two kinds of root systems—taproots and fibrous roots.

Root Study

1. My root is from a _____.

2. My root is colored _____.

3. My root is _____ cm long and _____ cm wide.

4. What are roots for?

Here is a drawing of my root.

Connecting Learning

1. Why do plants need roots?

2. Do all roots look alike? How are they the same? How are they different?

3. Would a tree have a taproot or fibrous roots? Why do you think so?

4. What kind of roots do many weeds have? Why might this be?

5. What are you wondering now?

THIS IS MY FLOWER

Topic
Flowers

Key Question
What is the function of a flower?

Learning Goal
Students will describe the parts of a flower.

Guiding Documents
Project 2061 Benchmark
- Tools such as thermometers, magnifiers, rulers, or balances often give more information about things than can be obtained by just observing things without their help.

NRC Standards
- Each plant or animal has different structures that serve different functions in growth, survival, and reproduction. For example, humans have distinct body structures for walking, holding, seeing, and talking.
- Ask a question about objects, organisms, and events in the environment.
- Employ simple equipment and tools to gather data and extend the senses.

*NCTM Standards 2000**
- Represent data using concrete objects, pictures, and graphs
- Count with understanding and recognize "how many" in sets of objects

Math
Graphing
  bar graph

Science
Life science
  botany
    flowers

Integrated Processes
Observing
Comparing and contrasting
Collecting and recording data

Materials
Flowers, one per student
Hand lenses, one per student
Contact paper (see *Management 2*)
Crayons
Construction paper frames (see *Management 3*)
Tape
Butcher paper
Student page

Background Information
A flower's function is to make seeds to reproduce the plant. The flower contains the reproductive parts of a flowering plant. Petals are not just to make the flower look pretty, but actually attract birds, bees, and other insects to the flower so that pollination can occur. When pollen from the stamen (male) lands on the stigma (female), a long pollen tube grows down the stalk of the pistil into the ovary. When ovules inside of the ovary are fertilized by pollen grains, they can develop into seeds. In some plants, the ovary then develops into a fruit that protects the seeds.

Management
1. Select flowers whose parts are easy to separate and observe such as daisies or lilies. Don't use flowers with too many petals. Each student needs one flower.
2. Each student will need and 8" x 10" piece of clear contact paper.
3. Make construction paper frames for students ahead of time. They should be the right size to frame an 8" x 5" piece of contact paper.

Procedure
1. Distribute the flowers and hand lenses. Have students carefully observe the parts of their flowers.
2. Hand out the student page and crayons and allow students to draw their flowers and record their observations.
3. Tell students to carefully take their flowers apart to observe the insides with the hand lens.

PRIMARILY PLANTS © 2005 AIMS Education Foundation

4. Discuss what students observe and talk about the purpose of flowers and the flower parts (see *Background Information*).
5. Give each student a piece of contact paper with the sticky side facing up. Instruct them to arrange their flowers' parts on one half of their papers. Show them how to carefully fold the paper in half so that it sticks together to form an 8" x 5" rectangle. Give students tape and their construction paper frames and have them attach the frames to their papers.
6. Make a large wall graph out of butcher paper to graph the number of petals on the flowers. Have students attach their flower frames to the graph in the appropriate places.

Connecting Learning
1. What did your flower look like? How was it like your classmates' flowers? How was it different?
2. What did you observe when you studied your flower with a hand lens?
3. What did you observe when you took your flower apart and studied it with a hand lens?
4. What purposes do the parts of the flower have?
5. Which number of petals was most common on the flowers in our class? How do you know?
6. What are you wondering now?

Extensions
1. Have students bring flowers from home. Describe and compare the flowers that come in. Make a graph of the colors or let students vote on their favorite color of flower.
2. Do a study on bees. Discuss how flowers attract bees. Explain how bees carry pollen to other flowers and what bees do with the pollen.

Curriculum Correlation
Art
Use flowers a models for artwork. Have students create their favorite flower in watercolors, crayons, chalk, torn construction paper, or a tissue paper collage.

Language Arts
Brainstorm words that begin with the letter *F* that describe flowers (fancy, flashy, full, fragrant, fuzzy, etc.).

Write two-word poetry about flowers:
> My flower,
> Looks soft,
> Smells sweet,
> Grows tall,
> Tickles me.

* *Reprinted with permission from Principles and Standards for School Mathematics, 2000 by the National Council of Teachers of Mathematics. All rights reserved.*

Key Question

What is the function of a flower?

Learning Goal

describe the parts of a flower.

THIS IS MY FLOWER

1. The colors in my flower are _____ _____

2. My flower has _____ petals.

3. My flower smells like _____ _____

4. Make a pollen print.

5. What are flowers for? _____
_____
_____
_____

Connecting Learning

1. What did your flower look like? How was it like your classmates' flowers? How was it different?

2. What did you observe when you studied your flower with a hand lens?

3. What did you observe when you took your flower apart and studied it with a hand lens?

4. What purposes do the parts of the flower have?

5. Which number of petals was most common on the flowers in our class? How do you know?

6. What are you wondering now?

Glossary

Algae:	a small plant that does not have roots, stems, leaves or flowers, but makes its own food
Bulb:	a large bud that is planted and grows to form roots, leaves, and flowers
Carbon Dioxide:	a gas in the air that is used by plants to make their food
Chlorophyll:	the green colored substance in plants that absorbs energy from sunlight
Cotyledon:	a seed leaf
Deciduous:	a kind of tree that loses its leaves every autumn
Dicotyledon:	also called a dicot, a plant whose seeds have two sections
Embryo:	the tiny plant within a seed
Flower:	the reproductive part of the plant that makes seeds and is often colorful
Fern:	a group of non-seed bearing plants that have roots, stems, and leaves
Fibrous Roots:	shallow roots that grow over a wide area
Frond:	a fern leaf that grows upward from the stem
Fruit:	the part of a plant that holds the seeds
Germinate:	when a seed starts to grow and produces a new plant
Leaf:	the flat, thin part of a plant that grows on the stem

Glossary

Monocotyledon: also called a monocot, a plant whose seeds have one section

Ovary: the bottom, rounded part of a pistil in which the ovules are located

Ovule: the female reproductive part of a plant that contains an egg

Petal: a part of a flower that often is brightly colored

Photosynthesis: the process in which green plants use light energy to make sugar from water and carbon dioxide

Pistil: female reproductive organ of a flower

Pollen: dust-like powder that carries the male genetic material of a plant

Pollination: transfer of pollen from the stamens to the pistils of flowers

Seed: the structure within a fruit that will grow into a new plant; it contains the embryo and a food supply

Seed Coat: the tough outside part found on many seeds

Spores: reproductive structure in some non-seed plants

Stamen: the pollen-producing male organ of a flower

Stigma: the very top of the pistil

Taproot: large central roots that grow deep into the soil

Tuber: the underground stem of a plant that stores starch

The AIMS Program

AIMS is the acronym for "Activities Integrating Mathematics and Science." Such integration enriches learning and makes it meaningful and holistic. AIMS began as a project of Fresno Pacific University to integrate the study of mathematics and science in grades K-9, but has since expanded to include language arts, social studies, and other disciplines.

AIMS is a continuing program of the non-profit AIMS Education Foundation. It had its inception in a National Science Foundation funded program whose purpose was to explore the effectiveness of integrating mathematics and science. The project directors in cooperation with 80 elementary classroom teachers devoted two years to a thorough field-testing of the results and implications of integration.

The approach met with such positive results that the decision was made to launch a program to create instructional materials incorporating this concept. Despite the fact that thoughtful educators have long recommended an integrative approach, very little appropriate material was available in 1981 when the project began. A series of writing projects have ensued, and today the AIMS Education Foundation is committed to continue the creation of new integrated activities on a permanent basis.

The AIMS program is funded through the sale of books, products, and staff development workshops and through proceeds from the Foundation's endowment. All net income from program and products flows into a trust fund administered by the AIMS Education Foundation. Use of these funds is restricted to support of research, development, and publication of new materials. Writers donate all their rights to the Foundation to support its on-going program. No royalties are paid to the writers.

The rationale for integration lies in the fact that science, mathematics, language arts, social studies, etc., are integrally interwoven in the real world from which it follows that they should be similarly treated in the classroom where we are preparing students to live in that world. Teachers who use the AIMS program give enthusiastic endorsement to the effectiveness of this approach.

Science encompasses the art of questioning, investigating, hypothesizing, discovering, and communicating. Mathematics is the language that provides clarity, objectivity, and understanding. The language arts provide us powerful tools of communication. Many of the major contemporary societal issues stem from advancements in science and must be studied in the context of the social sciences. Therefore, it is timely that all of us take seriously a more holistic mode of educating our students. This goal motivates all who are associated with the AIMS Program. We invite you to join us in this effort.

Meaningful integration of knowledge is a major recommendation coming from the nation's professional science and mathematics associations. The American Association for the Advancement of Science in *Science for All Americans* strongly recommends the integration of mathematics, science, and technology. The National Council of Teachers of Mathematics places strong emphasis on applications of mathematics such as are found in science investigations. AIMS is fully aligned with these recommendations.

Extensive field testing of AIMS investigations confirms these beneficial results:
1. Mathematics becomes more meaningful, hence more useful, when it is applied to situations that interest students.
2. The extent to which science is studied and understood is increased, with a significant economy of time, when mathematics and science are integrated.
3. There is improved quality of learning and retention, supporting the thesis that learning which is meaningful and relevant is more effective.
4. Motivation and involvement are increased dramatically as students investigate real-world situations and participate actively in the process.

We invite you to become part of this classroom teacher movement by using an integrated approach to learning and sharing any suggestions you may have. The AIMS Program welcomes you!

AIMS Education Foundation Programs

Practical proven strategies to improve student achievement

When you host an AIMS workshop for elementary and middle school educators, you will know your teachers are receiving effective usable training they can apply in their classrooms immediately.

Designed for teachers—AIMS Workshops:
- Correlate to your state standards;
- Address key topic areas, including math content, science content, problem solving, and process skills;
- Teach you how to use AIMS' effective hands-on approach;
- Provide practice of activity-based teaching;
- Address classroom management issues, higher-order thinking skills, and materials;
- Give you AIMS resources; and
- Offer college (graduate-level) credits for many courses.

Aligned to district and administrator needs—AIMS workshops offer:
- Flexible scheduling and grade span options;
- Custom (one-, two-, or three-day) workshops to meet specific schedule, topic and grade-span needs;
- Pre-packaged one-day workshops on most major topics—only $3,900 for up to 30 participants (includes all materials and expenses);
- Prepackaged *week-long* workshops (four- or five-day formats) for in-depth math and science training—only $12,300 for up to 30 participants (includes all materials and expenses);
- Sustained staff development, by scheduling workshops throughout the school year and including follow-up and assessment;
- Eligibility for funding under the Eisenhower Act and No Child Left Behind; and
- Affordable professional development—save when you schedule consecutive-day workshops.

University Credit—Correspondence Courses

AIMS offers correspondence courses through a partnership with Fresno Pacific University.
- Convenient distance-learning courses—you study at your own pace and schedule. No computer or Internet access required!

The tuition for each three-semester unit graduate-level course is $264 plus a materials fee.

The AIMS Instructional Leadership Program

This is an AIMS staff-development program seeking to prepare facilitators for leadership roles in science/math education in their home districts or regions. Upon successful completion of the program, trained facilitators become members of the AIMS Instructional Leadership Network, qualified to conduct AIMS workshops, teach AIMS in-service courses for college credit, and serve as AIMS consultants. Intensive training is provided in mathematics, science, process and thinking skills, workshop management, and other relevant topics.

Introducing AIMS Science Core Curriculum

Developed in alignment with your state standards, AIMS' Science Core Curriculum gives students the opportunity to build content knowledge, thinking skills, and fundamental science processes.
- *Each* grade specific module has been developed to extend the AIMS approach to full-year science programs.
- *Each* standards-based module includes math, reading, hands-on investigations, and assessments.

Like all AIMS resources these core modules are able to serve students at all stages of readiness, making these a great value across the grades served in your school.

For current information regarding the programs described above, please complete the following:

Information Request

Please send current information on the items checked:

____ *Basic Information Packet* on AIMS materials   ____ Hosting information for AIMS workshops
____ *AIMS Instructional Leadership Program*   ____ AIMS Science Core Curriculum

Name _____   Phone _____

Address_____
        Street   City   State   Zip

AIMS Magazine

YOUR K-9 MATH AND SCIENCE CLASSROOM ACTIVITIES RESOURCE

The AIMS Magazine is your source for standards-based, hands-on math and science investigations. Each issue is filled with teacher-friendly, ready-to-use activities that engage students in meaningful learning.

- *Four issues each year (fall, winter, spring, and summer).*

Current issue is shipped with all past issues within that volume.

| 1820 | Volume XX | 2005-2006 | $19.95 |
| 1821 | Volume XXI | 2006-2007 | $19.95 |

Two-Volume Combination

| M20507 | Volumes XX & XXI | 2005-2007 | $34.95 |

Back Volumes Available
Complete volumes available for purchase:

1802	Volume II	1987-1988	$19.95
1804	Volume IV	1989-1990	$19.95
1805	Volume V	1990-1991	$19.95
1807	Volume VII	1992-1993	$19.95
1808	Volume VIII	1993-1994	$19.95
1809	Volume IX	1994-1995	$19.95
1810	Volume X	1995-1996	$19.95
1811	Volume XI	1996-1997	$19.95
1812	Volume XII	1997-1998	$19.95
1813	Volume XIII	1998-1999	$19.95
1814	Volume XIV	1999-2000	$19.95
1815	Volume XV	2000-2001	$19.95
1816	Volume XVI	2001-2002	$19.95
1817	Volume XVII	2002-2003	$19.95
1818	Volume XVIII	2003-2004	$19.95
1819	Volume XIX	2004-2005	$35.00

Call today to order back volumes: 1.888.733.2467.

Call **1.888.733.2467** or go to **www.aimsedu.org**

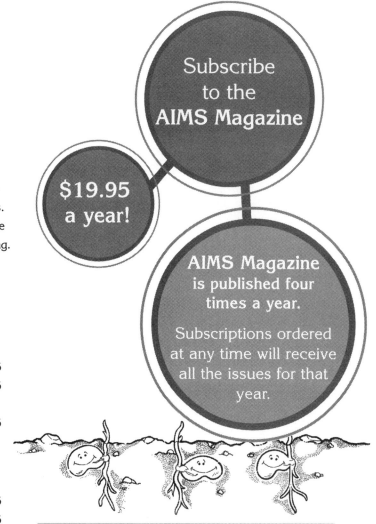

AIMS Online – www.aimsedu.org

For the latest on AIMS publications, tips, information, and promotional offers, check out AIMS on the web at www.aimsedu.org. Explore our activities, database, discover featured activities, and get information on our college courses and workshops, too.

AIMS News

While visiting the AIMS website, sign up for AIMS News, our FREE e-mail newsletter. Published semi-monthly, AIMS News brings you food for thought and subscriber-only savings and specials. Each issue delivers:

- Thought-provoking articles on curriculum and pedagogy;
- Information about our newest books and products; and
- Sample activities.

Sign up today!

AIMS Program Publications

Actions with Fractions, 4-9
Awesome Addition and Super Subtraction, 2-3
Bats Incredible! 2-4
Brick Layers II, 4-9
Chemistry Matters, 4-7
Counting on Coins, K-2
Cycles of Knowing and Growing, 1-3
Crazy about Cotton, 3-7
Critters, 2-5
Down to Earth, 5-9
Electrical Connections, 4-9
Exploring Environments, K-6
Fabulous Fractions, 3-6
Fall into Math and Science, K-1
Field Detectives, 3-6
Finding Your Bearings, 4-9
Floaters and Sinkers, 5-9
From Head to Toe, 5-9
Fun with Foods, 5-9
Glide into Winter with Math & Science, K-1
Gravity Rules! 5-12
Hardhatting in a Geo-World, 3-5
It's About Time, K-2
It Must Be A Bird, Pre-K-2
Jaw Breakers and Heart Thumpers, 3-5
Looking at Geometry, 6-9
Looking at Lines, 6-9
Machine Shop, 5-9
Magnificent Microworld Adventures, 5-9
Marvelous Multiplication and Dazzling Division, 4-5
Math + Science, A Solution, 5-9
Mostly Magnets, 2-8
Movie Math Mania, 6-9
Multiplication the Algebra Way, 4-8
Off the Wall Science, 3-9
Our Wonderful World, 5-9
Out of This World, 4-8
Overhead and Underfoot, 3-5
Paper Square Geometry:
   The Mathematics of Origami, 5-12
Puzzle Play, 4-8
Pieces and Patterns, 5-9
Popping With Power, 3-5
Positive vs. Negative, 6-9

Primarily Bears, K-6
Primarily Earth, K-3
Primarily Physics, K-3
Primarily Plants, K-3
Problem Solving: Just for the Fun of It! 4-9
Problem Solving: Just for the Fun of It! Book Two, 4-9
Proportional Reasoning, 6-9
Ray's Reflections, 4-8
Sense-Able Science, K-1
Soap Films and Bubbles, 4-9
Solve It! K-1: Problem-Solving Strategies, K-1
Solve It! 2nd: Problem-Solving Strategies, 2
Spatial Visualization, 4-9
Spills and Ripples, 5-12
Spring into Math and Science, K-1
The Amazing Circle, 4-9
The Budding Botanist, 3-6
The Sky's the Limit, 5-9
Through the Eyes of the Explorers, 5-9
Under Construction, K-2
Water Precious Water, 2-6
Weather Sense: Temperature, Air Pressure, and Wind, 4-5
Weather Sense: Moisture, 4-5
Winter Wonders, K-2

Spanish/English Editions*
Brinca de alegria hacia la Primavera con las
   Matemáticas y Ciencias, K-1
Cáete de gusto hacia el Otoño con las
   Matemáticas y Ciencias, K-1
Conexiones Eléctricas, 4-9
El Botanista Principiante, 3-6
Los Cinco Sentidos, K-1
Ositos Nada Más, K-6
Patine al Invierno con Matemáticas y Ciencias, K-1
Piezas y Diseños, 5-9
Primariamente Física, K-3
Primariamente Plantas, K-3
Principalmente Imanes, 2-8

* All Spanish/English Editions include student pages in Spanish and teacher and student pages in English.

Spanish Edition
Constructores II: Ingeniería Creativa Con Construcciones LEGO® 4-9
   The entire book is written in Spanish. English pages not included.

Other Science and Math Publications
Historical Connections in Mathematics, Vol. I, 5-9
Historical Connections in Mathematics, Vol. II, 5-9
Historical Connections in Mathematics, Vol. III, 5-9
Mathematicians are People, Too
Mathematicians are People, Too, Vol. II
What's Next, Volume 1, 4-12
What's Next, Volume 2, 4-12
What's Next, Volume 3, 4-12

For further information write to:
AIMS Education Foundation • P.O. Box 8120 • Fresno, California 93747-8120
www.aimsedu.org • Fax 559.255.6396

Duplication Rights

Standard Duplication Rights

Purchasers of AIMS activities (individually or in books and magazines) may make up to 200 copies of any portion of the purchased activities, provided these copies will be used for educational purposes and only at one school site.

Workshop or conference presenters may make one copy of a purchased activity for each participant, with a limit of five activities per workshop or conference session.

Standard duplication rights apply to activities received at workshops, free sample activities provided by AIMS, and activities received by conference participants.

All copies must bear the AIMS Education Foundation copyright information.

Unlimited Duplication Rights

To ensure compliance with copyright regulations, AIMS users may upgrade from standard to unlimited duplication rights. Such rights permit unlimited duplication of purchased activities (including revisions) for use at a given school site.

Activities received at workshops are eligible for upgrade from standard to unlimited duplication rights.

Free sample activities and activities received as a conference participant are not eligible for upgrade from standard to unlimited duplication rights.

Upgrade Fees

The fees for upgrading from standard to unlimited duplication rights are:
- $5 per activity per site,
- $25 per book per site, and
- $10 per magazine issue per site.

The cost of upgrading is shown in the following examples:
- activity: 5 activities x 5 sites x $5 = $125
- book: 10 books x 5 sites x $25 = $1250
- magazine issue: 1 issue x 5 sites x $10 = $50

Purchasing Unlimited Duplication Rights

To purchase unlimited duplication rights, please provide us the following:
1. The name of the individual responsible for coordinating the purchase of duplication rights.
2. The title of each book, activity, and magazine issue to be covered.
3. The number of school sites and name of each site for which rights are being purchased.
4. Payment (check, purchase order, credit card)

Requested duplication rights are automatically authorized with payment. The individual responsible for coordinating the purchase of duplication rights will be sent a certificate verifying the purchase.

Internet Use

Permission to make AIMS activities available on the Internet is determined on a case-by-case basis.

• P. O. Box 8120, Fresno, CA 93747-8120 •
• aimsed@aimsedu.org • www.aimsedu.org •
• 559.255.6396 (fax) • 888.733.2467 (toll free) •